T0276845

Handbook of Ferroelectrics

Handbook of Ferroelectrics

Edited by **Sharon Tatum**

New York

Published by NY Research Press,
23 West, 55th Street, Suite 816,
New York, NY 10019, USA
www.nyresearchpress.com

Handbook of Ferroelectrics
Edited by Sharon Tatum

© 2015 NY Research Press

International Standard Book Number: 978-1-63238-248-1 (Hardback)

This book contains information obtained from authentic and highly regarded sources. Copyright for all individual chapters remain with the respective authors as indicated. A wide variety of references are listed. Permission and sources are indicated; for detailed attributions, please refer to the permissions page. Reasonable efforts have been made to publish reliable data and information, but the authors, editors and publisher cannot assume any responsibility for the validity of all materials or the consequences of their use.

The publisher's policy is to use permanent paper from mills that operate a sustainable forestry policy. Furthermore, the publisher ensures that the text paper and cover boards used have met acceptable environmental accreditation standards.

Trademark Notice: Registered trademark of products or corporate names are used only for explanation and identification without intent to infringe.

Printed in the United States of America.

Contents

Preface   VII

Chapter 1   **Raman Scattering Study on the Phase Transition Dynamics of Ferroelectric Oxides**   1
Hiroki Taniguchi, Hiroki Moriwake, Toshirou Yagi and Mitsuru Itoh

Chapter 2   **Electromechanical Coupling Multiaxial Experimental and Micro-Constitutive Model Study of $Pb(Mg_{1/3}Nb_{2/3})O_3$-$0.32PbTiO_3$ Ferroelectric Single Crystal**   17
Wan Qiang, Chen Changqing and Shen Yapeng

Chapter 3   **Phase Transitions, Dielectric and Ferroelectric Properties of Lead-free NBT-BT Thin Films**   44
N. D. Scarisoreanu, R. Birjega, A. Andrei, M. Dinescu, F. Craciun and C. Galassi

Chapter 4   **'Universal' Synthesis of PZT (1-X)/X Submicrometric Structures Using Highly Stable Colloidal Dispersions: A Bottom-Up Approach**   62
A. Suárez-Gómez, J.M. Saniger-Blesa and F. Calderón-Piñar

Chapter 5   **Thin-Film Process Technology for Ferroelectric Application**   89
Koukou Suu

Chapter 6   **Ferroelectrics at the Nanoscale: A First Principle Approach**   120
Matías Núñez

Chapter 7   **Nanoscale Ferroelectric Films, Strips and Boxes**   147
Jeffrey F. Webb

Chapter 8   **The Influence of Vanadium Doping on the Physical and**
Electrical Properties of Non-Volatile Random Access Memory
Using the BTV, BLTV, and BNTV Oxide Thin Films   172
Kai-Huang Chen, Chien-Min Cheng, Sean Wu,
Chin-Hsiung Liao and Jen-Hwan Tsai

Chapter 9   **Emerging Applications of Ferroelectric Nanoparticles in**
Materials Technologies, Biology and Medicine   191
Yuriy Garbovskiy, Olena Zribi and Anatoliy Glushchenko

Chapter 10 **Photorefractive Effect in Ferroelectric Liquid Crystals**   214
Takeo Sasaki

Permissions

List of Contributors

Preface

In my initial years as a student, I used to run to the library at every possible instance to grab a book and learn something new. Books were my primary source of knowledge and I would not have come such a long way without all that I learnt from them. Thus, when I was approached to edit this book; I became understandably nostalgic. It was an absolute honor to be considered worthy of guiding the current generation as well as those to come. I put all my knowledge and hard work into making this book most beneficial for its readers.

Ferroelectricity is one of the most studied phenomena in the scientific community because of the vitality of ferroelectric materials in a broad spectrum of applications comprising of high dielectric constant capacitors, pyroelectric devices and transducers for medical diagnostic, piezoelectric sonars, electro-optic light valves, electromechanical transducers and ferroelectric random access memories. Ferroelectricity at nanoscale draws huge attention to the advancement of novel technologies. The need for ferroelectric systems with particular applications led to detailed research along with the betterment of processing and characterization techniques. This book provides an updated outlook of current research into ferroelectricity, covering several formulations, fabrication, properties, and theoretical topics.

I wish to thank my publisher for supporting me at every step. I would also like to thank all the authors who have contributed their researches in this book. I hope this book will be a valuable contribution to the progress of the field.

Editor

Raman Scattering Study on the Phase Transition Dynamics of Ferroelectric Oxides

Hiroki Taniguchi, Hiroki Moriwake,
Toshirou Yagi and Mitsuru Itoh

Additional information is available at the end of the chapter

1. Introduction

Ferroelectric phase transitions are conventionally divided into two types: an order-disorder and a displacive-type.[1] In the former one, which is frequently seen in hydrogen-bond-type ferroelectrics such as KH_2PO_4 (KDP), local dipole moments μs are randomly distributed between opposite directions in the paraelectric phase, leading to zero macroscopic net polarization P (= $\Sigma\mu$). A spontaneous polarization in the ferroelectric phase is then driven by their ordering through the ferroelectric phase transition. In the later type, in contrast, there are no dipole moments in the paraelectric phase. The spontaneous polarization in the ferroelectric phase stems from relative polar displacement between cationic and anionic sublattices. It is induced by freezing of a so-called ferroelectric "soft mode", which is known as a strongly anharmonic optical phonon mode at Γ-point in the Brillouin zone (BZ). The displacive-type transition is often found in the ferroelectric oxides as represented by perovskite-type compounds such as $PbTiO_3$. Since the ferroelectric oxides have been widely applied for the electronic devices, such as actuators, sensors, and memories, due to their chemical stability, the large spontaneous polarization, and the relatively high transition temperature, a better understanding of the soft mode behavior would be important from viewpoints of both application and fundamental sciences. Spectroscopic techniques employing light scattering, infrared absorption, and neutron scattering have been generally utilized for investigating the dynamics of the soft mode. Among them, Raman scattering has an advantage especially in a low-frequency region, which is significant to resolve the critical dynamics of the soft mode near the transition temperature. In the present review, we will discuss the soft mode behavior at the ferroelectric phase transition by referring the Raman scattering studies on $CdTiO_3$ and ^{18}O-substituted $SrTiO_3$. Furthermore, a microscopic origin of the soft mode is figured out from a viewpoint of local chemical bonds with the help of first-principles calculations.

2. Ferroelectric soft mode in the classical scheme

Since a concept of the soft mode was proposed, many theoretical and experimental studies have been conducted to clarify its dynamics.[2-4] In this section, we discuss the soft mode behavior in the classical scheme.

2.1. Theoretical background of the ferroelectric soft mode in the classical scheme

The potential of the soft mode is approximately described by accounting only a biquadratic phonon-phonon interaction as follows:

$$U_0 = \frac{1}{2}M(s,0)\omega_{\text{bare}}^2(s,0)Q^2(s,0) + \sum_\lambda \sum_k \frac{1}{2}J(s,0;\lambda,k)Q^2(s,0)Q^2(\lambda,k) \tag{1}$$

where $M(s,0)$, $\omega_{\text{bare}}(s,0)$, $Q(s,0)$ in the first term are effective mass, bare frequency, and displacement of the soft mode at the Γ-point. The $J(s,0;\lambda,k)$ in the second term denotes an interaction between the soft mode at the Γ-point and the optical phonon mode λ at the wave vector k, where the summation runs over all blanches and BZ. The equation can be transformed by a pseudo-harmonic approximation as

$$\begin{aligned}
U_0 &\approx \frac{1}{2}M(s,0)\omega_{\text{bare}}^2(s,0)Q^2(s,0) + \sum_\lambda \sum_k \frac{1}{2}J(s,0;\lambda,k)Q^2(s,0)\langle Q^2(\lambda,k)\rangle \\
&= \frac{1}{2}M(s,0)\left[\omega_{\text{bare}}^2(s,0) + \sum_\lambda \sum_k \frac{J(s,0;\lambda,k)}{M(s,0)}\langle Q^2(\lambda,k)\rangle\right]Q^2(s,0) \\
&\equiv \frac{1}{2}M(s,0)\omega_s^2 Q^2(s,0)
\end{aligned} \tag{2}$$

where the inside of the brackets is redefined by the soft mode frequency ω_s. In the classical regime, the mean square displacement $<Q^2(\lambda,k)>$ is proportional to $k_B T$, where k_B is a Boltzmann constant. The soft mode frequency is thus expressed as

$$\omega_s^2 \equiv \omega^2(s,0) = \omega_{\text{bare}}^2(s,0) + \sum_\lambda \sum_k \frac{J(s,0;\lambda,k)}{M(s,0)}\frac{k_B T}{M(\lambda,k)\omega^2(\lambda,k)} \tag{3}$$

Since ω_s is ideally zero at T_c of the second-order phase transition,

$$\omega_s^2 = \omega_{\text{bare}}^2(s,0) + \sum_\lambda \sum_k \frac{J(s,0;\lambda,k)}{M(s,0)}\frac{k_B T_c}{M(\lambda,k)\omega^2(\lambda,k)} = 0 \tag{4}$$

leading to

$$\omega_{\text{bare}}^2(s,0) = -\sum_{\lambda}\sum_{k} \frac{J(s,0;\lambda,k)}{M(s,0)} \frac{k_{\text{B}}T_{\text{c}}}{M(\lambda,k)\omega^2(\lambda,k)} \tag{5}$$

We thus obtain

$$\omega_s^2 = \sum_{\lambda}\sum_{k} \frac{J(s,0;\lambda,k)}{M(s,0)} \frac{k_{\text{B}}T}{M(\lambda,k)\omega^2(\lambda,k)} - \sum_{\lambda}\sum_{k} \frac{J(s,0;\lambda,k)}{M(s,0)} \frac{k_{\text{B}}T_{\text{c}}}{M(\lambda,k)\omega^2(\lambda,k)} \tag{6}$$

Defining C as

$$C \equiv \left[\sum_{\lambda}\sum_{k} \frac{J(s,0;\lambda,k)}{M(s,0)} \frac{k_{\text{B}}}{M(\lambda,k)\omega^2(\lambda,k)} \right]^{1/2} \tag{7}$$

it finally comes to Cochran's law

$$\omega_s = C(T - T_{\text{c}})^{1/2} \tag{8}$$

where

$$T_{\text{c}} \equiv -\omega_{\text{bare}}^2(s,0) \left[\sum_{\lambda}\sum_{k} \frac{J(s,0;\lambda,k)}{M(s,0)} \frac{k_{\text{B}}}{M(\lambda,k)\omega^2(\lambda,k)} \right]^{-1} \tag{9}$$

As seen in the above equation, the soft mode decreases its frequency with approaching T_c, and finally freezes to induce the ferroelectric phase transition. In the second-order phase transition, in particular, the displacement of the soft mode $Q(s,0)$ corresponds to the polar displacement in the ferroelectric phase. It should be noted that the bare frequency $\omega_{\text{bare}}(s,0)$ of the soft mode is assumed to be imaginary at zero-Kelvin. The softening of the soft mode connects to the divergent increase of the dielectric constant through Lyddane-Sachs-Teller (LST) relation,

$$\varepsilon' \propto \frac{1}{\omega_s^2} \tag{10}$$

2.2. Experimental observation of the ferroelectric soft mode in the classical scheme

The experimental observation of the typical soft mode behavior in the perovskite-type ferroelectric oxide, $CdTiO_3$, is presented in this section.

CdTiO$_3$ possesses an orthorhombic *Pnma* structure at room temperature due to $(a^+b^-b^-)$-type octahedral rotations in the Glazer's notation from the prototypical Cubic *Pm-3m* structure. The CdTiO$_3$ undergoes ferroelectric phase transition into the orthorhombic *Pna2$_1$* phase around 85 K.[5-7]

The confocal micro-Raman scattering is the one of useful techniques to investigate the soft mode dynamics in the displacive-type ferroelectric phase transition due to the following reason; In general, the complicated domain structure forms across the ferroelectric phase transition, where the principle axis of the crystal orients to various directions, which are allowed by the symmetry relation between the paraelectric and the ferroelectric phases. The size of the individual domain generally ranges over several nanometers to microns. When the domains are smaller than the radiated area of the incident laser, the observed Raman spectrum is composed of signals from differently oriented domains, leading to difficulty in the precise spectral analyses. With application of the confocal micro-Raman scattering, on the other hand, we can selectively observe the spectrum from the single domain region, because its spatial resolution reaches sub-microns not only for lateral but also depth directions.[8] Note that the Raman scattering can observe the soft mode only in the non-centrosymmetric phase due to the selection rule, therefore the critical dynamics of the soft mode is in principle investigated in the ferroelectric phase below T_c.

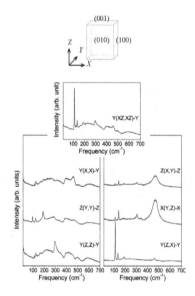

Figure 1. Raman spectra of CdTiO$_3$ at room temperature observed with several scattering geometries. See text for configuration notations, for instance Y(X,X)-Y.

Flux-grown colorless single crystals of CdTiO$_3$ with a rectangular solid shape were used for this study. The two samples have dimensions of approximately 0.3×0.2×0.1 mm^3 and

0.2×0.1×0.1 mm^3. Since the present samples were twinned, we carefully determined the directions of the axes in the observed area by checking the angular dependence of the Raman spectra with the confocal micro-Raman system, whose spatial resolution is around 1 μm (Fig. 1). Here we use laboratory coordinates X, Y, and Z, corresponding respectively to the crystallographic axes of [100], [010], and [001] in the paraelectric phase, to denote the Raman scattering geometries. For example, XZ(Y,Y)-Z-X means that the incident laser polarized parallel to the [010] direction penetrates the sample along the [101] direction, and the [010]-polarized scattered light is collected with the backscattering geometry. The scattered light was analyzed by a Jovin-Yvon triple monochromator T64000 with the subtractive dispersion mode, and was finally observed by a liquid-N$_2$-cooled CCD camera. The frequency resolution of the present experiment was better than 1 cm^{-1} The temperature of the sample was controlled by an Oxford microstat with a temperature stability of ±0.01 K.

Figure 2 shows the temperature dependence of the Raman spectrum of CdTiO$_3$, which is observed in Y(ZX,ZX)-Y scattering geometry with several temperatures from 85.0 K to 40.0 K. In the vicinity of $T_{c'}$ in particular, spectra were observed with the temperature intervals of 0.5 K. As shown in the figures, the soft mode is seen at around 60 cm^{-1} at 40 K. It gradually softens toward zero-frxuency as approaching $T_c \sim 85$ K. The spectral profile of the soft mode is analyzed by a damped-harmonic-oscillator (DHO) model;

$$I(\omega) = \frac{B\omega_0^2\Gamma\omega}{(\omega_0^2 - \omega^2)^2 + \Gamma^2\omega^2} \tag{11}$$

where B, ω_{0}, and Γ denote amplitude, harmonic frequency, and a damping constant of the soft mode, respectively. The soft mode frequency ω_s is defined here as

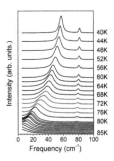

Figure 2. Temperature Dependence of the Raman spectrum of CdTiO$_3$ observed in the Y(ZX,ZX)-Y scattering geometry, and from 40K to 85 K on heating.

$$\omega_s^2 = \sqrt{\omega_0^2 - \Gamma^2} \tag{12}$$

The temperature dependence of the soft mode frequency and the damping constant, which are determined by the analyses, are plotted by solid circles and open squares in Fig. 3. Synchronizing with the softening of the soft mode, the damping constant is increased as approaching T_c.

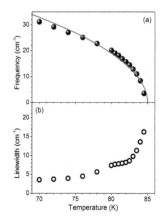

Figure 3. Temperature Dependencies of (a) the soft mode frequency and (b) the damping constant of CdTiO$_3$. The solid line in (a) is calculated by Cochran's law (see text).

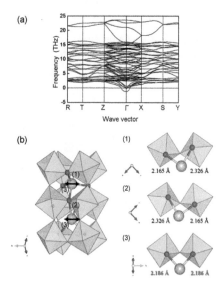

Figure 4. a) Phonon dispersion in CdTiO$_3$ obtained by the first-principles calculations. (b) The displacement pattern of the soft mode, and configurations of three O-Cd-O bonds along (1) [-101], (2) [101], and (3) [010] directions.

The temperature dependence of the soft mode frequency obeys the Cochran's low (Eq. 1.8) with $C = 8.5$ and $T_c = 85$ K as expressed by the solid curve in the figure, indicating the typical displacive-type ferroelectric phase transition on $CdTiO_3$.

The displacement pattern of the soft mode is obtained by first principle calculations. The calculations were conducted with a pseudopotential method based on a density functional perturbation theory with norm-conserving pseudopotentials, which was implemented in the CASTEP code.[9] Figure 4a presents phonon dispersion curves of $CdTiO_3$ in the paraelectric *Pnma* phase, which was obtained by the first-principle calculations. As seen in the figure, the soft mode, which is indicated by an imaginary frequency, is observed at Γ-point in Brillouin zone. According to the calculation, a displacement pattern of the Γ-point soft mode is composed of relative displacement of Ti with respective to the octahedra and an asymmetric stretching of O-Cd-O bond. It has been previously proposed that the hybridization between O-2p and empty Ti-3d orbitals triggers the non-centrosymmetric displacement of Ti thorough a second-order Jahn-Teller (SOJT) effect to induce the ferroelectricity as exemplified by $BaTiO_3$.[10] The SOJT effect, however, does not provide clear understanding on the role of *A*-site ion. In the case of $CdTiO_3$, in particular, the ferroelectricity disappears when the Cd is replaced by Ca, though the $CaTiO_3$ is an isomorph of $CdTiO_3$. Therefore, the ferroelectricity in $CdTiO_3$ can not be explained only by the *B*-O orbital hybridization. Here, we focus on the asymmetric stretching of O-Cd-O to elucidate the mechanism of ferroelectricity in $CdTiO_3$. The asymmetric stretching in the Γ-point soft mode is schematically illustrated by arrows in the right panel in Fig. 4b, where large and small spheres denote Cd and O ions, respectively. Ti and O ions that do not participate in the O-Cd-O bonds are omitted for simplicity. In a prototypical *Pm-3m* structure of the perovskite-type oxide, an *A*-site ion is surrounded by neighboring twelve O ions. In the $CdTiO_3$ of paraelectric *Pnma* structure, however, the number of closest oxygen ions for Cd ion is reduced to four due to the octahedral rotations. (Note that the octahedral rotations in the *Pnma* structure are different from each other according to the direction; the rotation is in-phase for a [010] direction whereas those for [101] and [-101] are anti-phase.) As a result, $CdTiO_3$ has three different O-Cd-O bonds along [-101], [101], and [010] directions, whose equilibrium structures are respectively denoted by (1), (2), and (3) in the Fig. 2(b). The result of first-principles calculations indicates that the soft mode displacement is inherently dominated by the asymmetric stretching of the O-Cd-O bond along the [010] direction denoted by (3). By comparing three equilibrium structures for O-Cd-O bonds presented in the right-hand side of Fig. 4b, it is found that the O-Cd-O bond along [010] direction is symmetric with the Cd-O bond lengths of 2.186 Å in the paraelectric *Pnma* phase, in contrast the other two are composed of non-equivalent bonds having 2.326 and 2.165 Å. Therefore, there is a remaining degree of freedom for further O-Cd-O off-center ordering due to the covalent bonding along the [010] direction. This remaining degree of freedom would be the origin of the ferroelectric soft mode.

3. A role of covalency in the ferroelectric soft mode

As indicated above, the covalency of the *A*-site ion plays an important role in the ferroelectricity in addition to that of *B*-site ion. The *A*-site substitution with an isovalent ion is instructive for the better understanding on the mechanism of the ferroelectricity in the perovskite-type oxides. This section is thus devoted to the effect of the isovalent Ca-substitution on the ferroelectricity of $CdTiO_3$. [11, 12]

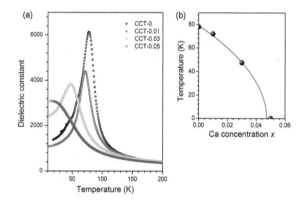

Figure 5. a) Dielectric constants of CCT-*x*. (b) Ca concentration dependence of the ferroelectric phase transition temperature T_c. Solid curve denotes Curie–Weiss fit of data points for CCT-0, 0.01, and 0.03.

Ceramics form of Ca-doped $CdTiO_3$, $Cd_{1-x}Ca_xTiO_3$ (CCT-*x*) with *x* = 0, 0.01, 0.03, 0.05, and 0.07, were fabricated by conventional solid state reactions from stoichiometric mixtures of CdO (3N), $CaCO_3$ (4N), and TiO_2 (3N). The mixtures were calcined during 24 h in air at 975 ~ 1175°C. Then the reground powders were pressed at 4 ton/cm², and sintered at 1200 ~ 1230°C. Dielectric measurements were performed by a LCZ meter.

Temperature dependencies of the dielectric constants in CCT-*x* are presented in Fig. 5a. As seen in the figure, the dielectric constant of CCT-0 divergently increases and culminates at the phase transition temperature T_c ~ 85 K. The T_c decreases with Ca doping, and the peak transforms into a low-temperature plateau at *x* = 0.05, indicating a quantum paraelectric state as discussed later. The Ca concentration dependence of T_c, which is estimated from the dielectric peaks, is shown in Fig. 5b. The fit curve with conventional Curie-Weiss law (a red line in the figure) indicates that the ferroelectricity in the CCT-*x* system is suppressed with a value of *x* larger than x_c = 0.047.

The temperature dependence of the soft mode frequency is presented in Fig. 6 as a function of *x*. The soft modes in CCT-*x* increase in frequency and intensity with distance from T_c on cooling, exhibiting typical behavior of the soft-mode-driven phase transition. According to the Lyddane–Sachs–Teller relationship (Eq. 1.10), the squared soft mode frequency ω_s^2 and the dielectric constant ε' are inversely proportional as mentioned in the preceding section. Therefore, if the dielectric properties of the CCT-*x* system are governed by the lattice dynamics, the soft mode frequency at the lowest temperature should decrease with increasing x because the dielectric constant near 0 K increases with x (see Fig. 5a). The Raman spectra at 4 K presented in the bottom panels in Fig. 6 show that the soft mode softens as x increases, indicating a displacive-type phase transition of the CCT-x system. This result suggests that the phase transition of CCT-x can be discussed in terms of lattice dynamics. Note that, the soft mode become observable at low temperature in CCT-0.05 and -0.07, although it does not undergo the ferroelectric phase transition within a finite temperature range. This is a typical characteristic of precursory softening of the soft mode in the quantum paraelectric state.[13, 14]

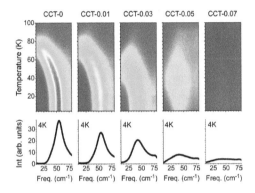

Figure 6. Contour plots for the temperature dependence of soft mode spectra in CCT-x (upper panels). Bottom panels show the low-frequency spectra observed at ~ 4 K.

Figure 7 presents the partial electronic density of states (p-DOS) for $CdTiO_3$ (left) and $CaTiO_3$ (right) obtained by the first-principles calculations, where the p-orbital of oxygen, the s-orbital of Cd and Ca, and the total DOS are denoted by oppositely hatched and blank areas, respectively. As shown in the figure, the s-orbital of Cd ion has strong hybridization with the p-orbital of oxygen. In marked contrast, the s-orbital of Ca has little hybridization, indicating larger A-O covalency in $CdTiO_3$ than $CaTiO_3$. This characteristic substantially agrees with the difference in electronegativity of Cd and Ca ions, where those in Cd and Ca are 1.7 and 1.0, respectively. A calculated charge density distributions around the O-Cd(Ca)-O bonds along the [010] direction are indicated in Fig. 8. As shown in the panels, larger charge density is observed between Cd and O ions in $CdTiO_3$, confirming its strong covalency. In contrast, Ca-O bond is nearly ionic. If the Cd(Ca) bonding is ionic, Cd(Ca) ion in the O-Cd(Ca)-O bond tends to locate a centric position. Therefore, Ca-substitution suppresses the freezing of the asymmetric stretching vibration with the off-centering of Cd(Ca) by decreasing the covalency. Since the asymmetric stretching of the O-Cd-O bond along [010] direction is suggested to be the origin of the soft mode, Ca-substitution results in the suppression of the softening of the soft mode. The present result confirms that the ferroelectric instability of CCT-x stems from A-site covalency.

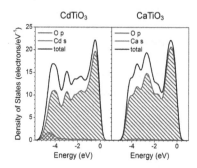

Figure 7. Partial electron density of states (p-DOS) for $CdTiO_3$ (left) and $CaTiO_3$ (right). Blue, red, and black curves represent p orbital of oxygen ions, s orbital of Cd and Ca ions, and total DOS.

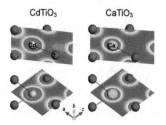

Figure 8. Cross sections of charge density around 4-coordinated Cd (left) and Ca (right) ions obtained by first principle calculations. Note that the upper and lower cross sections include O–Cd(Ca)–O bonds, which are involved in O–Cd(Ca)–O chains along the b and a direction, respectively.

4. Ferroelectric Soft Mode in the Quantum Scheme

It has been known that the quantum fluctuation plays a non-negligible role in the phase transition dynamics when the T_c goes down near 0 K, where the transition is suppressed though the dielectric permittivity reaches to several tens thousand. This effect is known as "quantum paraelectricity", which was first proposed as an origin of the giant dielectric plateau of $SrTiO_3$ at the low-temperature.[15] Twenty years later, it has been discovered that an isotope substitution with ^{18}O induces the ferroelectricity of $SrTiO_3$, attracting considerable attention of researchers.[16] In this section, the dynamics of quantum para/ferroelectric is discussed from a view point of the soft mode.

4.1. Theoretical background of the ferroelectric soft mode in the quantum scheme

In the classical case as mentioned before, the soft mode frequency can be described as Eq. 1.3. At the low-temperature, where the influence of the quantum fluctuation can not be ignored, the approximation $\langle Q^2(\lambda, k)\rangle \sim k_B T$ is not appropriate. Therefore, Eq. 1.3 should be modified with quantum statistic treatment as

$$\omega_s^2 = \omega_{\text{bare}}^2(s,0) + \sum_\lambda \sum_k \frac{J(s,0;\lambda,k)}{M(s,0)} \frac{\hbar}{2M(\lambda,k)\omega(\lambda,k)} \coth \frac{\hbar\omega(\lambda,k)}{2k_B T} \tag{13}$$

At the sufficiently low-temperature, where only the Γ-point soft mode that is the optical phonon with lowest energy can be thermally excited, any other optical phonons are not effective in the second term except for $\lambda = s$ at $k = 0$. The soft mode frequency can thus be expressed as

$$\omega_s^2 = \omega_{\text{bare}}^2(s,0) + \frac{J(s,0;s,0)}{2M^2(s,0)} \frac{\hbar}{\omega_s} \coth \frac{\hbar\omega_s}{2k_B T} \tag{14}$$

As in the classical case,

$$\omega_s^2 = \omega_{\text{bare}}^2(s,0) + \frac{J(s,0;s,0)}{2M^2(s,0)} \frac{\hbar}{\omega_s} \coth \frac{\hbar\omega_s}{2k_B T_c} = 0 \tag{15}$$

gives

$$\omega_{\text{bare}}^2(s,0) = -\frac{J(s,0;s,0)}{2M^2(s,0)} \frac{\hbar}{\omega_s} \coth \frac{\hbar\omega_s}{2k_B T_c} \tag{16}$$

We thus obtain the soft mode frequency at the low-temperature,

$$\omega_s = C' \left(\frac{T_1}{2} \coth \frac{T_1}{2T} - T_c' \right)^{1/2} \tag{17}$$

with

$$C' \equiv \left[\frac{J(s,0;s,0)}{M^2(s,0)} \frac{k_B}{\omega_s^2} \right]^{1/2}, \quad T_1 \equiv \frac{\hbar\omega_s}{k_B}, \quad T_c' \equiv \frac{T_1}{2} \coth \frac{T_1}{2T_c} \tag{18}$$

In this treatment, the temperature dependence of the squared soft mode frequency, which is inversely proportional to the dielectric permittivity, in no longer linear with respect to temperature, but saturated with the constant value near 0 K. The qualitative behavior of the soft mode in the quantum para/ferroelectrics is schematically illustrated as a function of T_1 with the fixed T'_c in Fig. 9.

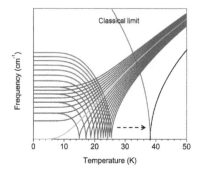

Figure 9. The schematic illustration describing the variation of the temperature dependence of the soft mode frequency as a function of T_1. See text for the detail.

The classical limit is also presented for comparison. Note that the value of T_1 is varied with constant intervals. As presented in the figure, the phase transition is completely suppressed when T_1 is sufficiently large, whereas it recovers as decreasing T_1. Interestingly, the T_1 dependence of the transition temperature becomes extremely sensitive when the transition temperature is close to 0K, suggesting that the quantum paraelectric-ferroelectric transition can be induced by subtle perturbation such as the isotope substitutions. [17]

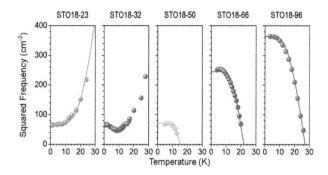

Figure 10. Temperature dependencies of the squared frequencies of the soft mode in STO18-x with various x. Solid lines in the figures denote temperature dependencies calculated by Eq. 3.7 with optimized control parameters (see text).

4.2. Experimental observation of the ferroelectric soft mode in the quantum scheme

The isotopically induced phase transition of the quantum paraelectric $SrTiO_3$ is a good example for the quantum paraelectric-ferroelectric phase transition driven by the soft mode.

Here we show the Raman scattering study on the soft mode in $SrTi(^{16}O_{1-x}{}^{18}O_x)_3$ (STO18-100x) as functions of temperature and the isotope substitution rate x. The experiments were performed with the Raman scattering geometries of X(YY)-X, where X, Y, and Z are denoted by the cubic coordination of $SrTiO_3$.[13,14,18]

Figure 10 presents the temperature dependence of the squared soft mode frequency in STO18-100x for 0.23, 0.32, 0.50 0.66, and 0.96, observed in the X(YY)-X scattering geometry, respectively. In $x = 0.50$, 0.66, and 0.96, the soft mode frequency is shown only for the ferroelectric phase. As indicated in the figure, the softening of the soft mode in STO18-23 saturates at the low-temperature region near 0 K, showing excellent agreement with the theory. Since the square of the soft mode frequency is inversely proportional to the dielectric permittivity as indicated by the LST-relation as mentioned before, it is clear that the dielectric plateau in $SrTiO_3$ stems from the soft mode dynamics. Note that the soft mode is nominally Raman inactive in the centrosymmetric structure as for the paraelectric phase of $SrTiO_3$ due to the selection rule. In the present case, however, the centrosymmetry is locally broken in the low-temperature region of STO18-100x ($x < x_c \sim 0.32$), leading to the observation of the soft mode even in the macroscopic centrosymmetry. The local non-centrosymmetric regions grow with ^{18}O-substitution to activate the soft mode spectrum even in the paraelectric phase as indicated in Fig. 11. A mechanism of such defect-induced Raman process and an expected spectral profile are discussed in detail in Ref. [13] and [14]. The date points for STO18-x ($x < x_c$) in Fig. 10 are obtained by the spectral fitting with the defect-induced Raman scattering model.

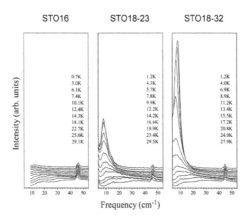

Figure 11. Temperature dependencies of the Raman spectra in STO18-x ($x = 0$ [STO16: the pure $SrTiO_3$], 0.23, and 0.32).

With the isotope substitution, softening of the soft mode is enhanced and the phase transition takes place above the critical concentration x_c as manifested by the hardening of the soft mode on cooling in the low temperature region. The transition temperature elevates as increasing x. The temperature dependencies of the soft modes in all samples except for that in

STO18-32 were examined by the generalized quantum Curie-Weiss law, which was proposed in Ref. 19, where the power of Eq. 3.5 is modified by $\gamma/2$;

$$\omega_s = C' \left(\frac{T_1}{2} \coth \frac{T_1}{2T} - T_c' \right)^{\gamma/2} \tag{19}$$

The γ is the one of critical exponents, which characterizes the critical behavior of the susceptibility. As seen in the figure, the plots are well reproduced by the fitting with the systematic variations of T_1 and γ as indicated in Fig. 12.

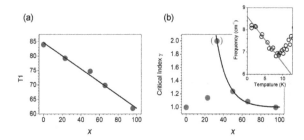

Figure 12. a) The x dependence of T_1 obtained by the fitting. The dashed line is a guide for the eye. (b) The x dependence of critical exponent γ. In STO18-23, -50, -66, -96, γ is determined by the fitting of the observed soft mode frequency with Eq. 3.7. In STO18-32, we directly determined $\gamma = 2$ from the linear temperature dependence of the soft mode frequency with T, as seen in the inset. The solid lines are eye guides.

In STO18-32, the strongly rounded soft mode behavior keeps us from fitting with Eq. 3.7. However, we can determine $\gamma = 2$ for STO18-32 from the obvious linear temperature dependence of the soft mode frequency, as seen in the inset of the panel (b). It should be noted here that $\gamma = 2$ corresponds to the theoretical value for the quantum ferroelectric phase transition [20–22], and is caused by the quantum mechanical noncommutativity between kinetic and potential energy. These results show experimentally that the quantum phase transition of STO18-x is an ideal soft mode-type transition driven by direct control of the quantum fluctuation with the ^{18}O-substitution. An origin of the strongly rounded softening was suggested in our previous study to be the nanoscopic phase coexistence between ferroelectric and paraelectric region. (See Ref. 18 for the detail)

5. Summary

In the present review, we overviewed the soft mode behavior in both classical and quantum schemes through Raman scattering experiment in the $CdTiO_3$ and the ^{18}O-substituted $SrTiO_3$. The analyses show excellent agreement qualitatively between the fundamental theory and experimental observations. The systematic studies with Raman scattering experiments and the first-principles calculations on Ca-substituted $CdTiO_3$ clarified that the covalency between constituent cation and oxygen plays an essential role in the origin of the soft mode. We hope the present review serves better understanding on the mechanism of displacive-type ferroelectric phase transition.

Author details

Hiroki Taniguchi[1]*, Hiroki Moriwake[2], Toshirou Yagi[3] and Mitsuru Itoh[1]

*Address all correspondence to:

1 Materials and Science Laboratory, Tokyo Institute of Technology, Yokohama, Japan

2 Nanostructures Research Laboratory, Japan Fine Ceramics Center, Nagoya, Japan

3 Research Institute for Electronic Science, Hokkaido University, Sapporo, Japan

References

[1] Line, M. E., & Glass, A. M. (1997). Principles and Applications of Ferroelectrics and Related Materials,. Claredon, Oxford,.

[2] Cochran, W. (1960). Crystal stability and the theory of ferroelectricity. *Adv. Phys.*, 9, 387.

[3] Cochran, W. (1961). Crystal stability and the theory of ferroelectricity part II. Piezo-electric crystals. *Adv. Phys.*, 10, 401 EOF-420 EOF.

[4] Scott, J. F. (1974). Soft-mode spectroscopy: Experimental studies of structural phase transitions. Rev. Mod. Phys. ., 46, 83 EOF-128 EOF.

[5] Shan, Y. J., Mori, H., Imoto, H., & Itoh, M. (2002). Ferroelectric Phase Transition in Perovskite Oxide $CdTiO_3$. *Ferroelectrics*, 381 EOF-386 EOF.

[6] Shan, Y. J., Mori, H., Tezuka, K., Imoto, H., & Itoh, M. (2003). Ferroelectric Phase Transition in $CdTiO_3$ Single Crystal. *Ferroelectrics*, 107 EOF-112 EOF.

[7] Moriwake, H., Kuwabara, A., Fisher, C. A. J., Taniguchi, H., Itoh, M., & Tanaka, I. (2011). First-principles calculations of lattice dynamics in $CdTiO_3$ and $CaTiO_3$: Phase stability and ferroelectricity. *Phys. Rev. B*, 84, 104.

[8] Taniguchi, H., Shan, Y. J., Mori, H., & Itoh, M. (2007). Critical soft-mode dynamics and unusual anticrossing in $CdTiO_3$ studied by Raman scattering. *Phys. Rev. B*, 76, 212103.

[9] Clark, S. J., Segall, M. D., Pickard, C. J., Hasnip, P. J., Probert, M. I. J., Refson, K., & Payne, M. C. (2005). First principles methods using CASTEP. *Z. Kristallogr*, 220, 567.

[10] Bersuker, I. B. (1995). Recent development of the vibronic theory of ferroelectricity:. *Ferroelectrics*, 164, 75.

[11] Taniguchi, H., Soon, H. P., Shimizu, T., Moriwake, H., Shan, Y. J., & Itoh, M. (2011). Mechanism for suppression of ferroelectricity in $Cd_{1-x}Ca_xTiO_3$. *Phys. Rev. B*, 84, 174106.

[12] Taniguchi, H., Soon, H. P., Moriwake, H., Shan, Y. J., & Itoh, M. (2012). Effect of Ca-Substitution on $CdTiO_3$ Studied by Raman Scattering and First Principles Calculations. *Ferroelectrics*, 268 EOF-273 EOF.

[13] Taniguchi, H., Takesada, M., Itoh, M., & Yagi, T. (2004). Effect of oxygen isotope exchange on ferroelectric microregion in $SrTiO_3$ studied by Raman scattering. *J. Phys. Soc. Jpn.*, 73, 3262 EOF-3265 EOF.

[14] Taniguchi, H., Takesada, M., Itoh, M., & Yagi, T. (2005). Isotope effect on the soft-mode dynamics of $SrTiO_3$ studied by Raman scattering. *Phys. Rev. B*, 72, 064111.

[15] Müller, K. A., Burkard, H., & Sr Ti, O. (1979). An intrinsic quantum paraelectric below 4 K. *Phys. Rev. B*, 19, 3593.

[16] Itoh, M., Wang, R., Inaguma, Y., Yamaguchi, Y., Shan, Y. J., & Nakamura, T. (1999). Ferroelectricity Induced by Oxygen Isotope Exchange in Strontium Titanate Perovskite. *Phys. Rev. Lett.*, 82, 3540 EOF-3543 EOF.

[17] Tokunaga, M., & Aikawa, Y. (2010). Isotope-Induced Ferroelectric Phase Transition in Strontium Titanate Only by a Decrease in Zero-Point Vibration Frequency. *J. Phys. Soc. Jpn.*, 79, 024707.

[18] Taniguchi, H., Itoh, M., & Yagi, T. (2007). Ideal Soft Mode-Type Quantum Phase Transition and Phase Coexistence at Quantum Critical Point in ^{18}O-Exchanged $SrTiO_3$. *Phys. Rev. Lett.*, 99, 017602 EOF.

[19] Dec, J., & Kleemann, W. (1998). From barrett to generalized quantum curie-weiss law. *Solid State Commun.*, 106, 695 EOF.

[20] Vojta, M. (2003). Quantum phase transitions. *Rep. Prog. Phys.*, 66, 2069.

[21] Sondhi, S. L., Girvin, S. M., Carini, J. P., & Shahar, D. (1997). Continuous quantum phase transitions. *Rev. Mod. Phys.*, 69, 315 EOF-333 EOF.

[22] Schneider, T., & Stoll, E. F. (1976). Quantum effects in an n-component vector model for structural phase transitions. *Phys. Rev. B*, 13, 1123 EOF-1130 EOF.

Chapter 2

Electromechanical Coupling Multiaxial Experimental and Micro-Constitutive Model Study of Pb(Mg$_{1/3}$Nb$_{2/3}$)O$_3$- 0.32PbTiO$_3$ Ferroelectric Single Crystal

Wan Qiang, Chen Changqing and Shen Yapeng

Additional information is available at the end of the chapter

1. Introduction

Compared to polycrystalline ferroelectric ceramics such as Pb(Zr$_{1/2}$Ti$_{1/2}$)-O$_3$ (PZT), domain engineered relaxor ferroelectric single crystals Pb(Zn$_{1/3}$Nb$_{2/3}$)O$_3$-xPbTiO$_3$ (PZN-xPT) and Pb(Mg$_{1/3}$Nb$_{2/3}$)O$_3$-xPbTiO$_3$ (PMN-xPT) show greatly enhanced electromechanical properties: the piezoelectric coefficient d_{33} and electrically induced strain of <001> oriented single crystals direction can respectively reach 2500pC/N and 1.7%, or even greater [1],[2]. PZN-xPT and PMN-xPT with outstanding properties are usually near the morphotropic phase boundary (MPB), separating rhombohedra (R) and tetragonal (T) phases [1], [4], [5]. For example, the MPB is located within a range of x=0.275-0.33 for PMN-xPT [6], [7]. Recent investigations have also shown that the MPB of PMN-xPT and PZN-xPT is actually in a multi-phase state at room temperature [8], [9]. e.g., the coexistence of rhombohedra (R) and monoclinic (M) phases at x=0.33 or rhombohedra and tetragonal phases at x=0.32. Other phases can also exist near MPB under stress and/or electric field loading. Experiment studies [10], [11], [12] and first-principles calculations [4], [13]reveal that two different homogeneous polarization rotation pathways are present between rhombohedra and tetragonal phases under an electric field. It is found that three kinds of intermediate monoclinic phases M$_A$, M$_B$ and M$_C$ are associated with the two pathways [9], [10]. Orthorhombic phase has also been observed between the rhombohedra and tetragonal phases when PMN-PT is loaded with strong electric field along the <110> direction [9],[14]. Although the mechanism underlying the high performance of these crystals is not completely clear, the existence of the intermediate phases and the associated phase transitions are believed to be one of the main reasons [1], [3], [4], [8], [10].

Note that PZN-PT and PMN-PT single crystals usually experience electric and/or mechanical loading during their in-service life. It has been shown that externally applied loading has significant effect on the properties of these crystals [1], [5]-[28]. A number of studies have focused on the loading induced behavior of <001> and <110> oriented anisotropic PZN-xPT and PMN-xPT crystals [1], [5], [9], [15]-[19]. It is has been shown loading in the form of electric field [1]-[21], and stress[21]-[26], can lead to polarization switching and phase transitions, which changes the crystal phase and domain structure of these single crystals, and hence dramatically alter their electromechanical properties. A mature level understanding of their responses to electrical, mechanical and temperature loading condition is thus essential to fulfill the applications of these crystals.

Rhombohedral phase of PZN-xPT($0<x<0.1$) and PMN-xPT($0<x<0.35$) exhibit excellent electromechanical properties along {001} and {110} orientations compared to the {111}(cubic cell reference) spontaneous polarization direction. A number of studies have focused on the loading induced behavior of {001}, {011} and {111} oriented PZN-xPT and PMN-xPT single crystals. It has been shown that loadings in the form of electric field and stress can lead to polarization rotation and phase transition, which change the crystal phase and domain structure of these single crystals, and hence dramatically alter their electromechanical properties. When an electric field is applied along the {001} direction of PZN-PT, polarization rotation occurs from {111} towards {001} via either M_A or M_B, depending the composition (e.g., the rotation is R-M_A-T for PMN-4.5PT). For {110}-oriented PMN-PT, polarization rotates from {111} to {110} via M_B when electric field is along {110} and ends up with an orthorhombic phase when the electric field exceeds a critical value. These single crystals can be polarized into a single domain state with {111} oriented electric field. Most available studies on the effect of bias stress on the crystal behavior focus only on uni-polar electric field loading. However, it has been shown that the uni-polar and bi-polar responses of these crystals can be very different.

A number of studies have focused on experimental, but the constitutive model of ferroelectric single is absent. Huber [29] and Bhattacharya[30], [31] et.al established a model based on the micromechanical method, but the simulation is not well compare to experimental result.

At present, a mature model to explain the stress-strain behavior of ferroelectric single is absent all along. In this study, electric field induced "butterfly" curves and polarization loops for a set of compressive bias stress of {001}, {011} and {111} poled PMN-0.32PT single crystals will be explored by systematical experiment study. The effects of the compressive bias stress on the material properties along these three crystallographic directions of PMN-0.32PT single crystals will be quantified. The underlying mechanisms for the observed feature will be explained in terms of phase transformation or domain switching,depending on the crystallographic direction.The stress-strain curves along <001> crystallographic direction of ferroelectric single crystals BaTiO$_3$ will be calculated in the first principle method to validate polarization rotation model. Finally, Based on the experimental phase transformation mechanism of ferroelectric single crystal, a constitutive model of ferroelectric single crystal is proposed based on micromechanical method. This constitutive model is facility and high computational efficiency.

2. Experimental methodology

At the test room temperature, PMN-0.32PT single crystals used in this study are of morphotropic composition, and in the rhombohedral phase, very close to MPB. The pseudo-
cubic {001}, {011} and {111} directions of these crystals are determined by x-ray
diffraction (XRD). Pellet-like specimens of dimensions 5×5×3mm^3 are then cut from
these crystals, with the normal of the 5×5mm^2 major specimen surfaces along the pseudo-cubic {001}, {011} or {111} direction. All specimens are electroded with silver on the
5×5mm major surfaces and poled along the {001}, {011} and {111} orientations (i.e., the
specimen thickness direction) under a field of 1.5kV/mm. Note that there are eight possible dipole orientations along the body diagonal directions of unpoled PMN-0.32PT single crystals (i.e., the <111> direction). When an electric poling field is applied to the
crystals along the {001} direction, a multi-domain structure can be produced, comprising
four degenerate states. For {011} direction poled single crystals, the number of degenerate states is two. The single crystals can be poled into a single domain state when they
are poled along the {111} direction. The states of {001}, {011} and {111} poled
PMN-0.32PT single crystals are sketched in Fig. 1, where solid arrows refer to the domain states induced by poling (also labeled as type 1 in Figs. 1b and 1c), dotted arrows
represent possible domains switched from type 1 domains upon loading or unloading,
and vice versa (also labeled as type 2,).

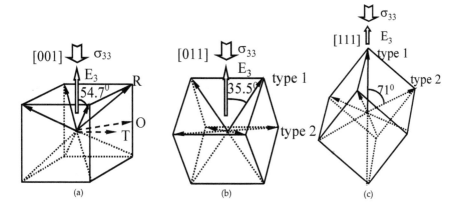

Figure 1. (a) Multidomain rhombohedral crystal obtained by poling along the {001} orientation. The solid line arrows
show possible directions of the polarization vector in a fully {001} poled rhombohedral crystal, the dashed line arrows
show possible directions of the polarization vector in orthorhombic and tetragonal phases crystal. (b) Two domain
rhombohedral crystal obtained by {011} poling. The solid line arrows indicate type 1 domains in a fully {011} poled
crystal and dashed line arrows indicate possible type 2 domains under compressive stress. (c) Monodomain rhombohedral crystal poled along the {111} orientation. The solid and dashed line arrows have similar means to (b). The hol-
low arrows in Fig. a ~ Fig. c show the electric field and stress loading. The small rhombohedral distortion is neglected
and all numbers and notations refer to a quasi-cubic unit cell.

Since the focus of this study is to explore the effect of bias stress on the electromechanical properties of PMN-0.32PT single crystals along different crystallographic directions, experimental setup is adapted from Ref. [31] (Fig.2) to allow simultaneously imposing uniaxial stress and electric field to the specimen along the thickness direction. Mechanical load is applied by a servo-hydraulic materials test system (MTS) and electric field is applied to the specimen using a high voltage power amplifier. Once the specimen is placed in the fixture, a compressive bias stress with magnitude of at least 0.4MPa is maintained throughout the test to ensure electrical contact. Stress controlled loading instead of displacement controlled loading is adopted during the test, so that the specimen is not clamped but is free to move longitudinally when electric field E_3 and mechanical stress σ_{33} loading are applied, where subscript 3 refers to the thickness direction of the samples, corresponding to the {001}, {011} and {111} direction of {001}, {011} and {111} oriented PMN-0.32PT single crystals, respectively. During test, polarization P_3 (or electric displacement) is measured using a modified Sawyer-Tower bridge, and the deformation (i.e., the strain) is monitored by two pairs of strain gauges (in total four strain gauges used) mounted on the four $5 \times 3mm^2$ surfaces: one pair placed on two opposite $5 \times 3mm^2$ surfaces is applied to measure the longitudinal normal strain ε_{33} along the {001}, {011} and {111} direction and another pair to measure the transverse normal strain ε_{11} in the direction perpendicular to the {001}, {011} and {111} direction respectively. Output of the strain gauges during deformation is recorded by a computer through a multiple-channel analog-to-digital (AD) converter.

The first set of tests is performed for electric field loading of triangular wave form of magnitude 0.5kV/mm and frequency 0.02Hz, free of stress loading. The low frequency is chosen to mimic quasistatic electric loading, which is of particular interest in this study [32]. Unless stated otherwise, this loading frequency for the electric field is used throughout the following test. The second set of tests consists of mechanical loading upon short circuited samples. The samples are compressed to −40MPa and unloaded to −0.4MPa at loading and unloading rate of 5MPa/min, followed by an electric field which is sufficiently large to remove the residual stress and strain to re-polarize the samples. In the third set of tests, triangular wave form electric field is applied to the samples which are simultaneously subject to co-axial constant compressive stress preload. The magnitude of the preload is varied from test to test and is in the range between 0 and -40MPa. Note that there is a time-dependent effect of the depolarization and strain responses under constant compressive stress. To minimize this effect, each electric field loading starts after a holding time of 150 seconds for a new stress preloading. It is found that three cycles of electric field loading and unloading are sufficient to produce stabilized response for each constant prestress, and the results for the last cycle are reported in the following.

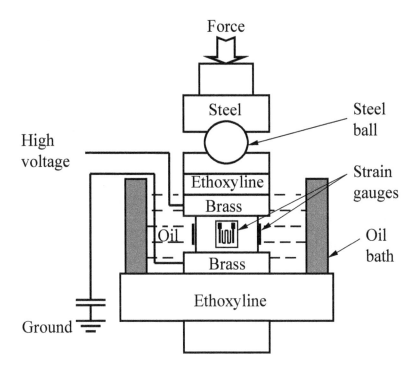

Figure 2. The sketch of experimental set

3. Experimental results and discussion

3.1. Crystallographic dependence of electric behavior and piezoelectric properties

The measured electric field induced polarization hysteresis loops and butterfly curves for {001}, {011} and {111} oriented poled PMN-0.32PT single crystals without stress loading are shown in Figs. 3a and 3b. The remnant polarizations P_r (defined as the polarization value at zero electric field), the coercive electric fields E_c (defined as the electric field value at zero polarization) and the piezoelectric coefficients d_{33} (defined as $d_{33}=\Delta\varepsilon_{33}/\Delta E_3$ where ΔE_3 is limited between –0.05 and +0.05 kV/mm, namely the slops of the $\varepsilon_{33}-E_3$ curves as the electric field passes through zero) depend strongly on the crystallographic orientation. From Figs. 3a and 3b, one can calculate that $P_{r_{[001]}}$, $P_{r_{[011]}}$ and $P_{r_{[111]}}$ are 0.247, 0.324 and 0.395C/m^2, $E_{c_{[001]}}$, $E_{c_{[011]}}$ and $E_{c_{[111]}}$ are 0.255, 0.298 and 0.216kV/mm, and $d_{33_{[001]}}$, $d_{33_{[011]}}$ and $d_{33_{[111]}}$ are 1828, 1049 and 200pC/N, respectively. It is noticed that there are eight possible polarization orientations along the pseudo-cubic {111} for un-poled rhombohedral PMN-0.32PT single crystals. Upon poling, the dipoles switch as close as possible to the applied electric field direction: For {001} poled crystals, there are four equivalent polar vectors along the {111} orientation, with an inclined angle of -54.7º from the poling field (Fig. 1a); For {011} poled crystals, there are two equivalent polar vectors along the {111} direction (labeled as type 1 in Fig. 1b); For {111} poled crystals, there is one polar vector along {111} (type 1 in Fig. 1c). According to the domain configurations in Fig. 1, the remnant polarizations $P_{r_{[001]}}$ and $P_{r_{[011]}}$ are approximately related to $P_{r_{[111]}}$ by $P_{r_{[001]}} = P_{r_{[111]}}/\sqrt{3}$ and $P_{r_{[011]}} = \sqrt{2}P_{r_{[111]}}/\sqrt{3}$, respectively. By taking the measured value for $P_{r_{[111]}}$ (0.395C/m²), $P_{r_{[001]}}$ and $P_{r_{[011]}}$ are predicted to be 0.228 and 0.322C/m², respectively, which are very close to the measured ones ($P_{r_{[001]}}$ =0.227C/m² and $P_{r_{[011]}}$=0.324C/m²). Therefore, the measured results are consistent with the domain configurations shown in Fig. 1.

It is also seen from Figs. 3a and 3b that the coercive field is lowest for the {111} oriented crystals, and becomes successively higher for {001} and {110} orientations. This is same to Ref.25 except for {110} orientation. This trend of coercive field is due to two reasons: One is due to reorientation driving force being proportional to the component of electric field aligned with the rhombohedral direction; The other one is due to the domain switching process. In {111}-oriented PMN-0.32PT single crystals, there are two types of domains (shown as type 1 and type 2 in Fig.1c). When the electric field is decreased from 0.5kV/mm to –0.05kV/mm, the strain first decreases linearly (see Fig.3b). When the electric field is decreased further, the type 1 domain switches to type 2 domains, leading to abrupt displacement change. When the electric field exceeds the coercive field –0.216kV/mm, the type 2 domains switch back to type 1 domain, recovering the deformation. In type 2 domain state, three equivalent polar vectors with an angle of 71º from the {111} direction can coexist and are separated by domain walls across which the normal components of electric displacement and displacement jump are zero. Ideally, this type of domain walls has no associated local stress or electric field. So the existing of type 2 domain state and the largest component of electric field along polarization direction induces the lowest coercive field in {111} orientation poled crystal. In the {011} orientation crystals, there are also two types of domains (i.e., type 1 and type 2 domain in Fig. 1b), the domain switching process is similar to {111}-poled crystals. In the type 2 domain state, however, the four possible polar vectors are perpendicular to the applied electric field (Fig. 1b). It is thus difficult to switch type 2 domain to type 1 domain only by applying electric field. Both the type 2 domain state and the smallest component of E contribute to the largest coercive field in {011} oriented crystals. In {001} oriented crystals, there will be no associated local stress or electric field, similar to type 2 domain state of {111} poled crystals. This feature again renders domain switching easy. On the other hand, the domain structure of {001} poled crystals is stable [4] and the component of electric field is smaller, giving the coercive field higher than that of {111} poled crystals but lower than that of {011} oriented crystals (see, Figs. 3a and 3b).

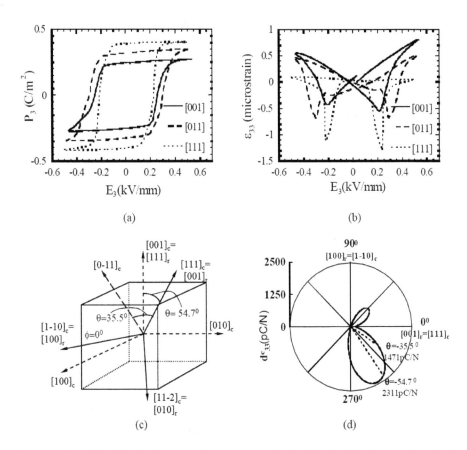

Figure 3. Electric field induced polarization and strain responses for {001}, {011} and {111}-oriented crystals of PMN-0.32PT: (a) P_3–E_3 curves and (b) ε_{33}–E_3 curves. (c) Relationship between rhombohedral (solid lines) and cubic (dashed lines) coordinate systems. Subscripts r and c denote directions with respect to the rhombohedral and cubic coordinate systems, respectively. (d) Orientational dependence of piezoelectric coefficient d^*_{33} of PMN-0.32PT in the polar plane.

Some researchers attribute the high piezoelectric coefficients along {001} and {011} oriented ferroelectric single crystals to the engineered domain state [4, 11]. It has also reported that, however, the piezoelectric coefficient along the {001} direction of single crystals with mono-domain structure is comparable to that of crystals with multi-domain structure [17], implying the origins of the high piezoelectric constants of PMN-0.32PT single crystals may not be due to the engineered domain state. Instead, it could be due to the effect of crystal lattice properties. To further explore this issue, we follow Ref. [17] to calculate the piezoelectric co-efficients d^*_{ij} along an arbitrary direction in a mono-domain crystal. For a direction defined by the Euler angles (φ, θ, ψ) (see Fig. 3c), d^*_{ij} is related to d_{ij} (measured along the principal crystallographic axes) by,

$$d^*_{33}(\phi,\theta) = d_{33}\cos^3\theta + (d_{15} + d_{31})\cos\theta\sin^2\theta - d_{22}\cos\phi\sin^3\theta(\cos^2\phi - 3\sin^2\phi) \qquad (1)$$

With d_{33}=200pC/N taken from our measured value and d_{15}=4100, d_{31}=−90 and d_{22}=1340pC/N, $d^*_{33[111]}(0, \theta)$ for PMN-0.32PT single crystals can be calculated and is shown in Fig. 3d. Note that in the case of ϕ=0 θ=-35.5º and -54.7º correspond to {011} and {001} direction, respectively. $d^*_{33[011]}$=1471pC/N and $d^*_{33[001]}$=2311pC/N can be inferred from Figs. 3d, which are not far away from our measured values ($d_{33[011]}$=1049 pC/N and $d_{33[001]}$=1828 pC/N). The small discrepancy between the predictions and measurements is believed to be due to the fact that the predictions are based on ideal mono-domain single crystals while the material parameters used in Eq. (1) are actually from less ideal mono-domain crystals (in fact, they are more or less multi-domain crystals). Nevertheless, we can conclude that the dominant contribution to the large {001} and {011} piezoelectric response should be the crystal anisotropy other than engineered domain state.

3.2. Crystallographic dependence of stress induced strain and polarization responses

Figure 4 shows the measured $\sigma_{33} - \varepsilon_{33}$ and $\sigma_{33} - P_3$ curves for {001}, {011} and {111} oriented short circuited samples. The results indicate obvious crystallographic anisotropy in stress inducing responses. From domain switching viewpoint, there should be no significant deformation in the thickness dimension of {001} poled crystal samples under stress loading, except for the elastic strain. However the longitudinal strain of {001} is contractive with a maximum magnitude of 0.3% under −40MPa. The contractive strain of {001} oriented crystals is about twice as much as those of {011} (about −0.18%) and {111} (about −0.17%) oriented crystals under a loading of −40MPa (Fig. 4a). This abnormal behavior of {001}-oriented crystals lies in that the mechanism underlying the stress induced response of {001}-oriented crystals is the R to O and T phase transition (see, Fig. 1a) rather than domain switching for {011} and {111} oriented crystals: the lattice distortion due to phase transition induces large deformation in the thickness dimension of {001} poled crystals. In {011} and {111}- poled crystals, on the contrary, the type 2 multi-domain state induced by compressive stress is the stable and preferred state (Figs. 1b and 1c), and can form more easily by domain switching than phase transformation. Therefore, the stress induced strain and polarization curves in {011} and {111} orientation crystals are similar to those of ferroelectric polycrystals of which domain switching is also the dominant deformation mechanism (Figs. 4a and 4b). During the unloading of {001} oriented crystals (corresponding to the returning to R phase of the unstable O phase), the $\sigma_{33} - \varepsilon_{33}$ and $\sigma_{33} - P_3$ curves show obvious nonlinear behavior. Note that the remnant strain and polarization at the end of unloading are attributed to the stable T phase which does not switches back to the R phase. For {011} and {111}-oriented crystals, however, there is only domain switching (i.e., switching between type 1 and type 2 domains) and no phase transformation occurs. During unloading, the $\sigma_{33}-\varepsilon_{33}$ curves and $\sigma_{33}-P_3$ curves of {011} and {111} show linear response since almost no domain switches back (Figs. 4a and 4b).

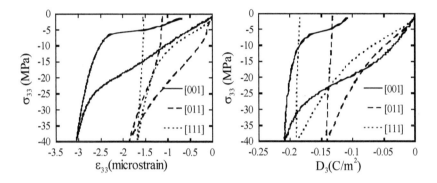

Figure 4. Compressive loading and unloading stress cycles induced polarization and strain responses for {001}, {011} and {111}-oriented crystals of PMN-0.32PT: (a) $\sigma_{33}-\varepsilon_{33}$ curves and (b) $\sigma_{33}-P_3$ curves.

The stress cycles of {001} poled crystals can be explained by the polarization vector rotation mechanism sketched in Fig. 5 as follows. In general two most possible mechanisms responsible for the observed features of {001} poled crystals in Figs.4 are domain switching and polarization rotation associated phase transformation. Recall that the PMN-0.32PT single crystals considered here are in the R phase close to MPB (although in reality there could also be M or T phase in these crystals near MPB [8], [9] the R phase is nevertheless the dominant phase). It is noted that that, upon application of a field along the <001> poling axis of R phase domain engineered PMN-0.32PT single crystals, only four of the eight polarization orientations are possible, i.e., <111>, <$\bar{1}$11>, <1$\bar{1}$1>, and <$\bar{1}\bar{1}$1>. Since the <001> components of these four polar vectors are completely equivalent, each domain wall cannot move under an external electric field along the <001> direction owing to the equivalent domain wall energies [32]. In other words, no ferroelectric domain switching is possible. Equally, no ferroelastic domain switching is expected when a compressive stress is applied along the <001> axis. (For PMN-PT systems with orthorhombic 4O <001> domain engineered structure [33], 60° switching from <011> to <110> would be driven by an applied stress. But, this is not what we are considering). Such facts partly rationalize our hypothesis that the underlying mechanism associated with the observed behavior of the R-phased PMN-0.32PT single crystals considered in this paper is phase transformation. Nevertheless, it should be emphasized that since no in situ diffraction observation is conducted to properly identify the stress induced phase transformation in PMN-0.32PT single crystals, discussions in this paper on phase transition are solely inferred from the measured stress induced strain and polarization curves.

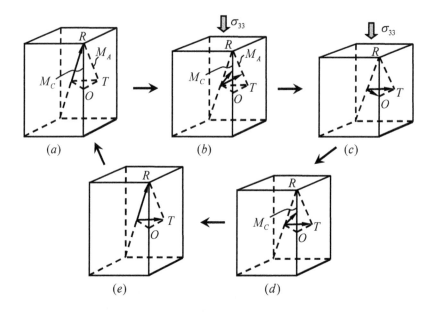

Figure 5. Schematic drawing of the polarization rotation from R to T and O phases due to <001> direction compression: (a) the initial state, (b) polarization starts to rotate from R to O and T via two pathways M_A and M_C, (c) R phase completely transformed, (d) O phase starts to switch back to R phase upon unloading, (e) O phase completely switched back to R phase. The states (a) to (e) correspond to the stress levels marked in Figs. 2(a) and 2(b). At the initial poled state (a), PMN-0.32PT possesses four equivalent <111>-oriented polarizations vector close to <001> direction though only one polarization rotation processes is shown here.

Fig. 5 illustrates the polarization rotation mechanism of PMN-0.32PT single crystals under stress loading and unloading cycle. The polarization vector states shown in Figs. 5(a)-5(e) correspond to the stress levels marked on the curves in Figs. 4 for the -40MPa stress cycle. In the initial poled state, PMN-0.32PT single crystals possess four equivalent <111> polarizations, with only one shown in Fig. 5a for the sake of clarity. Upon loading, polarization vector starts to rotate from R to O and T phases through intermediate phases M_A and M_C, when the compressive stress exceeds about 15MPa in magnitude (Fig. 5(b)). This gives a mixture of R and T phases and remarkable augment in the polarization and strain change around σ_{33} =-20MPa (Figs. 4). When the presence of an electric field along <110> direction, polarization rotation only occurs from R to O under compression along <001> direction. In the absence of electric field along <110>, however, almost all the polarization vectors may switch to T and O phases when the compression exceed 30MPa in magnitude (Fig. 5(c)). When the magnitude of the compressive loading is further increased, there is no more polarization rotation and the PMN-0.32PT single crystal shows linear response in Figs. 4. It is noted that O phase is usually unstable (its free energy balance between R and T phases depends on the electric and mechanical loading history [9], [14]). Upon unloading, the O phase switches back to R phase when the compression lower than -10MPa (Figs. 4, and 5(d)). On the other hand T phase is a stable phase. It will not switch back to R phase upon unloading, leading to remnant strain of about -0.07% for and remnant polarization of about -0.13C/m^2 at the end of the -40MPa stress cycle (Figs. 4(a) and 4(b), Fig. 5(e)). After the stress cycle, a unipolar electric field of 0.5kV/mm in magnitude is applied to re-pole the single crystal to its initial state, i.e., the polarization vector completely switches back. As a result, the remnant strain and polarization diminish to zero at the end of the re-polarization process (Figs. 5(a) and 5(e)).

3.3. Crystallographic dependence of electric field induced behavior at constant bias compressive stress

The $\varepsilon_{33}-E_3$ "butterfly" curves and P_3-E_3 hysteresis loops of {001}, {011} and {111} poled crystals under different constant compressive bias stresses are shown in Fig. 6. Dependence of the electric coercive E_c, remnant polarization P_r, dielectric permittivity χ_{33}, piezoelectric coefficient d_{33} and aggregate strain $\Delta \varepsilon$ for {001}, {011} and {111} oriented crystals on the compressive bias stress is summarized in Fig. 7. The aggregate strain $\Delta \varepsilon$ is defined as the difference between the maximum and minimum strain for a complete responsive butterfly curves. Similar to d_{33}, χ_{33} are calculated by $\Delta P_3 / \Delta E_3$ where ΔP_3 is the polarization difference between –0.05 and +0.05kV/mm. The calculated χ_{33} within such a small field range is almost equal to the slope of P_3-E_3 hysteresis loops when the electric field passes through zero. The calculated χ_{33} includes both the reversible (intrinsic dielectric property) and irreversible (extrinsic domain switching and phase transformation related property) contributions of the material, which is generally higher than the permittivity measured by a dynamic method [32].

The influence of the preloaded compressive stress on the aggregate strain $\Delta \varepsilon$, remnant polarization P_r and piezoelectric coefficient d_{33} seems to be similar for {001}, {011} and {111} oriented crystals: The remnant polarization P_r decreases with increasing the magnitude of the compressive prestress; The aggregate strain $\Delta \varepsilon$ and piezoelectric coefficient d_{33} first increase and then decrease with the magnitude of the prestress increasing. As suggested earlier, however, the underlying mechanisms for the electromechanical behavior of {001} {011}, and {111} oriented crystals are different. The change of the dielectric permittivity χ_{33} {001} oriented crystals under compressive stress is similar to that of χ_{33} induced by temperature: near the phase transformation temperature there is a peak in the χ_{33} curves, As is shown in Fig. 7c, there is a small peak near –6MPa and a big peak near –20MPa in χ_{33} curves for {001} oriented crystals. This may be due to R-M (near –6MPa) and M-R and M-O phase transformation (near –20MPa). On the other hand, there is no obvious peak in χ_{33} curves for {011} and {111}-oriented crystals (Fig.7c). The change of χ_{33} curves induced by compressive stress and the aforementioned stress induced strain and polarization responses are consistent with the hypothesis that there is phase transformation for {001}-orientated crystals under compressive stress.

As suggested by Fu et al. [7], for {001} oriented crystal origins of large aggregate strain and high piezoelectric coefficient at a moderate compressive bias stress (i.e., around -20MPa for the single crystals considered here) may be attributed to the stress induced intermediate states between rhombohedral and tetragonal phases. Under a bias stress of about -20MPa, PMN-0.32PT single crystals, after a phase transformation, are in a state of monoclinic phase which has a larger c/a (c and a are lattice parameters) and a smaller polarization component along the field direction than those of rhombohedral phase [20], implying electric field induced greater aggregate strain and piezoelectric coefficient (Figs. 7d and 7e). With the magnitude of the compressive bias stress increased further (i.e., σ_{33}=-30 and -40MPa), the {001} oriented single crystals are in the state of a mixture of orthorhombic and tetragonal phases. Although O and T phases also have large c to a ratio, the presence of a large compression has an opposite effect (i.e., preventing larger deformation induced by electric field) and results in small aggregate strain (Fig.7e).

As is noted, the polarization rotation introduced by compression gives rise to θ in $d_{33}^{*}(\phi, \theta)$ in the range between -54.7° and -90° (the angle between loading and polarization direction of T or O phase) under -20MPa, and results in larger d_{33}^{*} in accordance with Eq. (1) (see in Fig. 7d). When the magnitude of the compressive bias stress increases further, θ approaches 90° and d_{33}^{*} becomes smaller. Meanwhile, the remnant polarization P_r and coercive field E_c decrease monotonically with the applied compressive bias stress because of the decreased component polarization along the {001} direction under compression (Fig.7b).

Under zero stress, the initial state of {011}-oriented crystals is of multi-domain with two equivalent polarization directions (Fig.1b). For {111}-oriented crystals, the initial state is of mono-domain with polarization direction in the {111} direction (Fig.1c). When the applied electric field decreases from 0.5kV/mm to $-E_c$, the strain and polarization decrease linearly. Upon approaching $-E_c$, domain state switches to four polar domain state for {011} oriented crystals (type 2 in Fig. 1b) and three polar domain state for {111} oriented crystals (type 2 in Fig. 1c), giving the jumps in strain and polarization responses. When electric field exceeds $-E_c$, domain complete second switching, from four polar domain state to two polar domain state for {011} oriented crystals and from three polar domain state to mono-domain state for {111} oriented crystals. When the electric field reaches -0.5kV/mm, the strain and polarization change linearly again (Fig. 7).

Note that compressive stress can induce domain switching in {011} and {111}-orientated crystals. When compressive stress is superimposed on the samples, the shapes of $\varepsilon_{33} - E_3$ and $P_3 - E_3$ curves are different from those under the zero stress state (Fig.6c-f). Due to compression inducing depolarization, the remnant polarization decreases in {011} and {111} oriented crystals (Fig. 7b). A low compressive stress (e.g., -5MPa for {011}, -7MPa for {111}) can lead to type 1 to type 2 domain switching. As a result, more domains take part in switching during the electric field loading cycle and larger aggregate strain$\Delta\varepsilon$ than under zero compressive stress is observed. This partly explains the observed features in Fig. 7e.

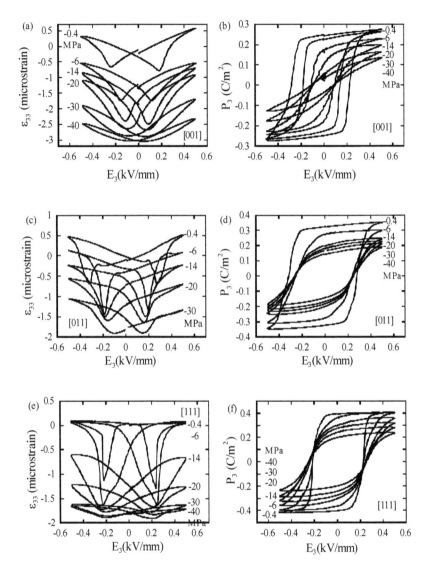

Figure.6 Electric field induced $_{33}$ E_3

Figure 6. Electric field induced $\varepsilon_{33}-E_3$ and P_3-E_3 curves at different compressive bias stresses: (a) and (b) for {001}-oriented; (c) and (d) for {011}-oriented, (e) and (f) for {111}-oriented.

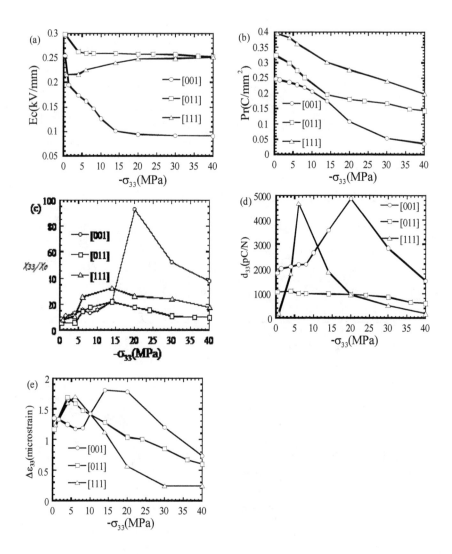

Figure 7. Effect of compressive bias stress on (a) the coercive field E_c, (b) remnant polarization P_r, (c) piezoelectric constant d_{33}, (d) relative dielectric constant X_{33}/X_0, and (e) and aggregate strain $\Delta\varepsilon_{33}$.

4. The first principle calculation of stress-strain

4.1. Calculation methodology

PMN-PT and BaTiO$_3$ have the similar ABO$_3$ structure, so they have the similar ferroelectric properties. There is only Ti^{4+} particle in B site of BaTiO$_3$, however, there is not only Ti^{4+} but also minim Mn^{4+} and Ni^{2+} particle in B site of PMN-PT. So the single cell of BaTiO$_3$ is convenient in calculation and it keeps the similar ferroelectric to PMN-PT.

In this paper, we calculate a single cell of BaTiO3, the single cell should be the smallest periodic reduplicate cell, it includes one Ti^{4+} particle, three O^{2-} particles and one Ba^{2+} particle (Fig.8). It is suggested that the initialized state of BaTiO$_3$ ferroelectric single crystal is R phase after {001} oriented polarization. Refer to literature [34], the crystal lattice constant of

R phase BaTiO$_3$ ferroelectric single crystal is 4.001A, the coordinate of particle in single cell is shown in table 1.The loading along {001} direction is carried out through application increasing strain by degrees. Stress and other parameters under each strain level is calculated by VASP. Calculation under each strain level include two steps. Firstly, the particle coordinate of single cell under each strain level is calcluated by first principle molecular dynamic method. Secondly, Stress and other parameters are obtained through relaxation that is based on the first result. Each increment of strain in this paper is 0.5%, untill 4%.

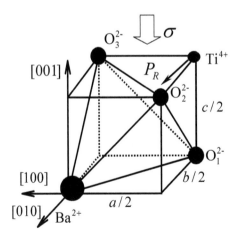

Figure 8. The cell calculation model of BaTiO$_3$ in R phase

4.2. Discussion

In order to validate that the method a mentioned in this paper is correct, we calculate the elasticity constant C_{33} of T phase BaTiO$_3$ using the method mentioned before. Fig.9 shows that the C_{33} is 180GPa, which is approach to the experimental value 189± 8Gpa reported in refer[34]. This means that the method used in this paper is trusty.

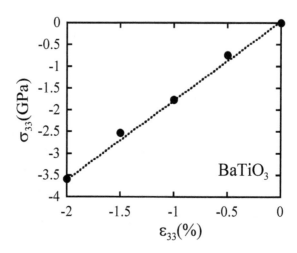

Figure 9. The stress-train curve of BaTiO$_3$ ferrorlctric single crystal obtained by the method amentioned in this paper, the {001} oriented BaTiO$_3$ ferroelectric single crystal is in T phase.

Stress-strain curve of {001} orientated R phase BaTiO$_3$ calculated in first principle method is shown in Fig.10(a). Fig.10(a) shows that the result has simlar nonlinear behavior to the experimental result of PMN-0.32PT during the loading that sketched in Fig.4(a), namely, there is obvious "a,b,c" steps during loading. We know that the nonlinear behavior of PMN-0.32PT shown in Fig.4(a)should be polarization rotation (R→M→O and R→M→T). The R→M→O is corresponding to the processing of polarization vector P_R switching to P_O, which is shown in Fig.10(b). table 2 is calculated coordinate of particle in BaTiO$_3$, the coordinate is correspond to point A in fig.10(a). Table 3 is coordinate of particle in O phase BaTiO$_3$ ferroelectric single crystal from refer [35]. Data in table 2 are equal to those in table 3, which indicates that "A" point in fig.10(a) should be O phase. With increasing strain, the coordinates of particle are unchangeable after "A", this indicates BaTiO$_3$ ferroelectric single crystal is stable in O phase after "A". From the first principle calculation, we testify that the PMN-0.32PT ferroelectric single crystal undergoes polarization rotation(R→M→O), this prove the polarization rotation model is reasonable.

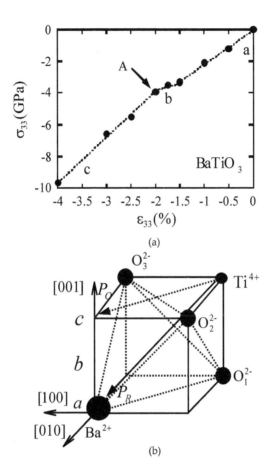

Figure 10. (a)The stress-strain curve of {001} oriented BaTiO$_3$ ferroelectric single crystal calculated in first principle calculation, the scatter point is calculation value and the broken line fit the scatter point.(b) The sketch map of polarization rotation during the calculation. The "a,b,c" letter in (a) is correspond to the "a,b,c" letter in (b), the letter indicates different polarization states during loading.

5. The viscoplastic model of ferroelection single

5.1. The viscoplastic model

From the experimental analysis, it is known that the <001>-oriented PMN-0.32PT single crystal undergo the R→M→T and R→M→O phase transformation under the compression. In order to establish a compact model and keep the essence of experiment, R→O phase transformation of PMN-0.32PT is considered in this study. Polarization rotation cause bulk deformation and shear deformation associated with the slip planes. It is assumed that the corresponding deformation associated with the slip planes is shear dominated, a feature similar to that of the multi-slip system of the crystal plasticity. This similarity renders possible to use crystal plasticity models to describe the transformation deformation of PMN-0.32PT single crystal.

The poled ferroelectric single is four domain state, the polarization vector is along <111>, <$\bar{1}$11>, <1$\bar{1}$1> and <$\bar{1}\bar{1}$1> respectively. In this study, it is assumed that the polarization vector of R phase can switch to the polarization vector of O phase, such as <111> vector switches to <001>, the vector also can switch back from O phase to R phase. The polarization rotation has an analogy to crystal plasticity slip, so we suggest there is eight slip systems in PMN-0.32PT ferroelectric single crystal. According to the crystalgraphic theory, we suggest ferroelectric single crystal has possible 8 variants; transformation of variants can be characterized by the habit planes illustrated in Fig.11. where n denotes the unit normal to the habit plane and s refers to the direction of transformation.

In general, a criterion (i.e., the phase transformation criterion) exists for the phase transformation of ferroelectric single crystal, and the material is assumed to undergo phase transformation when at least one of the 8 variants satisfies the phase transformation criterion. This is detailed as follows. Upon loading, the condition to produce R→O polarization rotation on a specified habit plane is that the driving force G of that plane reaches the critical value G_{0O}. The driving force is composed of the chemical driving force G_{chem} and mechanical driving force G_{mech}[36]

$$G = G_{chem} + G_{mech} = G_{0O} \qquad (2)$$

A similar condition holds for reverse transformation from O→R polarization rotation with a critical value G_{0A}

$$G = G_{chem} + G_{mech} = G_{0A} \qquad (3)$$

In Eqs. (2) and (3), the mechanical driving force can be expressed by

$$G_{mech}^r = \tau^a \gamma^* + E^a P^* \qquad (4)$$

Where τ^a and E^a are the resolved stress on the "a" transformation system, and γ^* and P^* denote the associated transformation strain and transformation polarization. Following the crystal theory of plasticity, the resolved stress τ^a of the variant "a" is related to the stress tensor σ_{ij} and the Schmid factor a_{ij}^a by

Electromechanical Coupling Multiaxial Experimental and Micro-Constitutive Model Study of
Pb(Mg$_{1/3}$Nb$_{2/3}$)O$_3$- 0.32PbTiO$_3$ Ferroelectric Single Crystal

35

$$\tau^a = \sum_{i,j=1}^{3} \alpha_{ij}^a \sigma_{ij} \tag{5}$$

where the Schmid factor is defined as follow

$$\alpha_{ij}^a = \frac{1}{2}(m_i^a n_j^a + m_j^a n_i^a) \tag{6}$$

with m_i and n_i being the unit normal to the habit plane and shear direction to the variant "a", respectively. The resolved stress E^a of the variant "a" is related to the electrical field tensor E_i

$$E^a = \sum_{i=1}^{3} E_i s_i \tag{7}$$

The chemical driving force in Eqs. (2) and (3) is assumed to be a linear function of the temperature

$$G_{chem} = \beta(T - T_0) \tag{8}$$

where T and T_0 denote the temperature in the single crystal and the equilibrium temperature respectively, and β is the stress-temperature coefficient. Note that the equilibrium temperature T_0 is defined as the average of the starting temperature of O phase transformation and that of R phase transformation, that is

$$T_0 = \frac{1}{2}(O_s + T_s) \tag{9}$$

When applying the rate independent crystal theory based model, Eqs. (2) and (3), to simulate the behavior of ferroelectric single crystals, one of the most computationally consuming tasks is to determine the set of instantaneously active transformation systems among the 8 possible variants at crystal level. This determination is usually achieved by an iterative procedure and must be carried out at each loading step, requiring extensive computation. Note that in the crystal theory of plasticity, a similar problem exists whilst in a rate dependent viscoplastic version of crystal theory of plasticity, determination of the set of active transformation systems is not necessary. As a result, computation effort can be reduced significantly. Following this idea, a viscoplastic version of Eqs. (21) and (32) are proposed and employed in this study. In the viscoplastic crystal model for PMN-0.32PT ferroelectric single crystals all transformation systems are assumed to be instantaneously active of varying extent, which is governed by a rate dependent viscoplastic law. In this paper, the phase transformation of variant "a" is assumed to comply with the following power law of viscoplasticity,

$$\dot{f}^a = \dot{f}_0 \left| \frac{G^a}{G_0} \right|^{\frac{1}{m}-1} \frac{G^a}{G_0} \left| \frac{c}{c_0} \right|^{1/k} \tag{10}$$

where $\dot{f}^a$ is phase fraction transformation rate of variant "a", $\dot{f}_0$ is reference phase fraction transformation rate, G^a is the driving force of variant "a", G_0 is the critical driving force, exponents k and m are material parameters dictating the rate effect, c depends upon the phase fraction, c_0 the reference value at initial state.

The phase fraction of variant "a" is calculated as,

$$\xi^a = f^a / f_0 \tag{11}$$

Summation over all possible variants provides the phase fraction for the whole ferroelectric single crystal, that is

$$\xi = \sum_{a=1}^{8} \xi^a \tag{12}$$

The incremental form of Eq. (12) can be written as

$$d\xi^a = df^a / f_0 \tag{13}$$

The transformation strain tensor ε_{ij}^{tr}, which is associated with df^a, can be obtained as

$$d\varepsilon_{ij}^{tr} = \sum_{r=1}^{8} \alpha_{ij}^a \gamma^* df^a \tag{14}$$

where α_{ij} is Schmid factor. $d\varepsilon_{ij}^{tr}$ is increment of phase train. γ^* is the maximal strain transformation during loading. df^a is increment of phase fraction.

Elastic strain and electrical field induced strain during loading can expressed as

$$d\varepsilon_{ij}^{e} = C_{ijkl}d\sigma_{ij} \qquad d\varepsilon_{ij}^{E} = d_{kij}dE_{k} \tag{15}$$

Increment of the total strain $d\varepsilon_{ij}$ can then expressed as the sum of an elastic component $d\varepsilon_{ij}^{e}$, a electrical field induced component $d\varepsilon_{ij}^{E}$ and a transformation induced component $d\varepsilon_{ij}^{tr}$,

$$d\varepsilon_{ij} = d\varepsilon_{ij}^{e} + d\varepsilon_{ij}^{E} + d\varepsilon_{ij}^{tr} \tag{16}$$

5.2. Numerical results

To validate the model, the corresponding calculation result is compared to experimental stress-strain curve. The material parameters used in this study are as follow [37, 38]. Material parameters of R phase are C_{11}^{R}=9.2GPa, C_{12}^{R}=10.3GPa, C_{44}^{R}=6.9GPa. γ^{*}=0.0033, G_{0R}=0.5073MJ/m^3, β=0.004MPa/K, $1/m$=11, $1/k$=1.0. T=403, T_0=298K. Material parameters of O phase are C_{11}^{O}=38GPa, C_{12}^{O}=40GPaC$_{44}^{O}$=28GPa. G_{0O}=0.276MJ/m^3 β=0.039MPa/K, $1/m$=11.5, $1/k$=1.1. Fig.12 shows the model predicted and experimental measured response. Result in fig.8 shows that stress-strain response of PMN-0.32PT can be predicted by the developed constitutive model, with quantitative agreement.

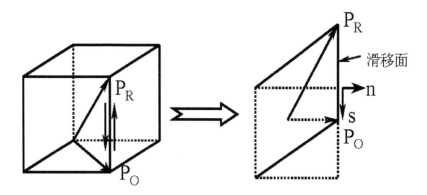

Figure 11. The comparison between ferroelectric phase transformation and plastic slip of single crystal, n is direction normal to slip surface, s is direction along phase transformation.

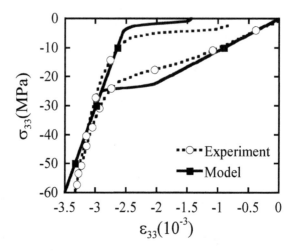

Figure 12. The comparison between experimental and simulation stress-strain curve along <001> direction of ferro-electric single crystal.

	100	010	001
Ba^{2+}	0	0	0
O$_1$ $^{2-}$	0.5133	0.5133	0.0192
O$_2$ $^{2-}$	0.5133	0.0192	0.5133
O$_3$ $^{2-}$	0.0192	0.5133	0.5133
Ti^{4+}	0.489	0.489	0.489

Table 1. The coordinate of particle in R phase BaTiO$_3$ ferroelectric single crystal cell [34]

	100	010	001
Ba^{2+}	0	0	0
O$_1$ $^{2-}$	0.514	0.514	0.01
O$_2$ $^{2-}$	0.514	0.016	0.521
O$_3$ $^{2-}$	0.016	0.521	0.499
Ti^{4+}	0.4846	0.4846	0.499

Table 2. The calculated coordinate that is correspond to point A in fig.6(a)

	100	010	001
Ba^{2+}	0	0	0
O$_1$$^{2-}$	0.5144	0.5144	0
O$_2$$^{2-}$	0.5	0.0162	0.523
O$_3$$^{2-}$	0.0162	0.523	0.5
Ti^{4+}	0.4857	0.4857	0.5

Table 3. The coordinate of particle in O phase BaTiO$_3$ ferroelectric single crystal cell from refer.

6. Conclusion and discussion

In this paper, Stress induced strain and polarization, and electric field induced "butter-fly" curves and polarization loops for a set of compressive bias stress for {001}, {011} and {111} poled PMN-0.32PT single crystals are experimentally explored. Obtained results indicate that high piezoelectric responses of PMN-0.32PT single crystals are controlled by the anisotropy of the crystals and the multi-domain structure (i.e., engineered domain structure) has a relatively minor effect. Analysis shows that in all three directions the electric field induced aggregate strain $\Delta\varepsilon$ and piezoelectric constant d_{33} increase with increasing the magnitude of the compressive bias stress. However, when the magnitude of the compressive bias stress is further increased the electric field induce $\Delta\varepsilon$ and d_{33} decrease. As a result, an optimized compressive bias stress exists for the purpose of enhancing the electromechanical properties of {001} and {111} oriented PMN-0.32PT single crystals. These results have apparent importance in the design of actuators and sensors using PMN-0.32PT single crystals. It is found that the observed stress induced strain and polarization in {001}-oriented PMN-0.32PT can be described by a polarization rotation mechanism, i.e., polarization rotates from rhombohedral (R) to orthohombic (O) and tetragonal (T) phases through the intermediate Monoclinic (M) phase during loading, and O to R transition during unloading. However, domain switching is believed to be the main mechanism dictating the electromechanical behavior of {011} and {111} oriented PMN-0.32PT single crystals. polarization rotation model is developed to explain the observed behaviors of PMN-0.32PT. The stress-strain curve along <001> crystallographic direction of ferroelectric single crystal BaTiO$_3$ is calculated with the first principle method. Obtained results show that the R→M→O polarization rotation (phase transformation) takes place in rhombohedral BaTiO$_3$ ferroelectric single crystals under compression, which is consistent with the polarization rotation model. Based on the polarization rotation model, a constitutive model of PMN-0.32PT is proposed based on micromechanical model. It is shown that the developed model can faithfully capture the key characteristic of the observed constitutive behavior of <001> oriented PMN-0.32PT.

Acknowledgments

The authors are grateful for the financial supported by the Natural Science Foundation of China (No. 10425210,10802081) and the Ministry of Education of China.

Author details

Wan Qiang[1], Chen Changqing[2] and Shen Yapeng[2]

1 Institute of Structural Mechanics, China Academy of Engineering Physics, Mianyang, Sichuan, China

2 Key Laboratory, School of Astronautics and Aeronautics, Xi'an Jiaotong University, Xi'an, China

References

[1] Park S. Shrout TR. Ultrahigh strain and piezoelectric behavior in relaxor based ferroelectric single crystals. Journal of Applied Physics 1997:82,1804-1812.

[2] Service RF. Shape-changing Crystals Get Shifter. Science 1997:275,1878-1880.

[3] Liu SF. Park S. Shrout TR. Cross LE. Electric field dependence of piezoelectric properties for rhombohedral 0.955Pb($Zn_{1/3}Nb_{2/3}$)O_3– 0.045PbTiO$_3$ single crystals Journal of Applied Physics 1999:85,2810-2815.

[4] Fu H. Cohen RE. Polarization Rotation Mechanism for Ultrahigh Electromechanical Response in Single-crystal Pizeoelectrics. Nature(London) 2000:403,281-283.

[5] Choi SW. Shrout TR. Jang SJ. Bhalla AS. Morphotropic phase boundary in Pb(Mg1/3Nb2/3)O3-PbTiO3 system. Mater. Lett. 1989:8(6-7),253–255

[6] Luo HS. Xu GS. et. Compositional Homogeneity and Electrical Properties of Lead Magnesium Niobate Titanate Single Crystals Grown by a Modified Bridgman Technique. Jpn. J. Appl. Phys. 2000:39,5581-5585.

[7] Sreemoolanadhan H. Dielectric ceramics in the BaO-Ln2O3-5TiO2 composition. Ferroelectrics. 1996:189,43-46.

[8] Xu GS. Luo HS. Xu HQ. Third ferroelectric phase in PMNT single crystals near the morphotropic phase boundary composition. Phys. Rev. B, 2001:64(2),020102-020105.

[9] Viehland D. Li JF. Amin A. Electromechanical and elastic isotropy in the (011) plane of 0.7Pb($Mg_{1/3}Nb_{2/3}$)O_3-0.3PbTiO$_3$ crystals: Inhomogeneous shearing of polarization. Journal of Applied Physics 2002:92,3985-3990.

[10] Noheda B. Zhong Z. Cox DE. Shirane G. Park SE. Electric-field-induced phase transitions in rhombohedral Pb(Zn$_{1/3}$Nb$_{2/3}$)$_{1-x}$Ti$_x$O$_3$. Physics Review B. 2002:65,224101-224104.

[11] Ye ZG. Noheda B. Dong M. et al. Monoclinic phase in the relaxor-based piezoelectric/ ferroelectric Pb(Mg$_{1/3}$Nb$_{2/3}$)O$_3$-PbTiO$_3$ system. Physics Review B 2001:64,184114-184117.

[12] Ge WW. Luo CT. Zhang QH. Devreugd CP. et al. Ultrahigh electromechanical response in (1–x)(Na0.5Bi0.5)TiO3-xBaTiO3 single-crystals via polarization extension. Journal of Applied Physics 2012:111, 093508-093511.

[13] Xing ZP. Xu K. Dai G. Li JF. Viehland D. Giant magnetoelectric torque effect and multicoupling in two phases ferromagnetic/piezoelectric system. Journal of Applied Physics 2011:110,104510-104514

[14] Tu CS. Tsai CL. Schmidt VH. Luo HS. Dielectric, hypersonic, and domain anomalies of Pb(Mg$_{1/3}$Nb$_{2/3}$)O$_3$-xPbTiO$_3$ single crystals. Journal of Applied Physics 2001:89(12), 7908-7916.

[15] Farokhipoor S. Noheda B. Conduction through 71° Domain Walls in BiFeO$_3$ Thin Films. Physics Review Letter 2011:107,127601-127605.

[16] Topolov VY. Cao H. Viehland D. Correlation between non-180° domain structures in (1–x)PbA1/3Nb2/3O3–xPbTiO3 single crystals (A = Mg or Zn) under an applied (001) electric field. Journal of Applied Physics 2007:102,024103-024106

[17] Christelle J. Li JF. Viehland D. Investigation of polarization switching in (001)c, (110)c, and (111)c oriented Pb(Zn$_{1/3}$Nb$_{2/3}$)O$_3$–4.5%PbTiO$_3$ crystals. Journal of Applied Physics 2004:95,5671-5675.

[18] Cao H. Li JF. Viehland D. Electric-field-induced orthorhombic to monoclinic MB phase transition in [111] electric field cooled Pb(Mg$_{1/3}$Nb$_{2/3}$O$_3$)–30%PbTiO$_3$ crystals. Journal of Applied Physics 2006:100,084102-084106

[19] Chen KP. Zhang XW. Luo HS. Electric-field-induced phase transition in <001>-oriented Pb(Mg$_{1/3}$Nb$_{2/3}$)O$_3$-PbTiO$_3$ single crystals. Journal of Physics: Condensed Matter 2002:14 (29),571-576.

[20] Tu CS. Tsai CL. Chen LF. Luo HS. Dielectric properties of relaxor ferroelectric Pb(Mg$_{1/3}$Nb$_{2/3}$)O$_3$-xPbTiO$_3$ single crystals. Ferroelectrics 2001:261(1-4),831-836.

[21] Catalan G. Janssens A. Rispens G. Polar Domains in Lead Titanate Films under Tensile Strain. Physics Review Letter 2006:96,127602-127606.

[22] Viehland D. Powers J. Ewart L. Ferroelastic switching and elastic nonlinearity in <001>-oriented Pb(Mg$_{1/3}$Nb$_{2/3}$)O$_3$–PbTiO$_3$ and Pb(Zn$_{1/3}$Nb$_{2/3}$)O$_3$–PbTiO$_3$ crystals. Journal of Applied Physics. 2000:88,4907-4912

[23]   Viehland D. Li JF. Investigations of electrostrictive $Pb(Zn_{1/3}Nb_{2/3})O_3$–$PbTiO_3$ ceramics under high-power drive conditions: Importance of compositional fluctuations on residual hysteresis. Journal of Applied Physics 2001:89,1826-1829.

[24]   Viehland D. Powers J. Effect of uniaxial stress on the electromechanical properties of $0.7Pb(Mg_{1/3}Nb_{2/3})O_3$–$0.3PbTiO_3$ crystals and ceramics. Journal of Applied Physics 2001:89,1820-1824.

[25]   Viehland D. Ewart L. Powers J. et al. Stress dependence of the electromechanical properties of <001> -oriented $Pb(Mg_{1/3}Nb_{2/3})O_3$–$PbTiO_3$ crystals: Performance advantages and limitations. Journal of Applied Physics 2001:90,2479-2484

[26]   Viehland D. Li JF. Anhysteretic field-induced rhombhohedral to orthorhombic transformation in <110> -oriented $0.7Pb(Mg_{1/3}Nb_{2/3})O_3$–$0.3PbTiO_3$ crystals. Journal of Applied Physics 2002:92,7690-7696

[27]   Noheda B. Cox DE. Shirane G. Park SE. Polarization Rotation via a Monoclinic Phase in the Piezoelectric 92% $PbZn_{1/3}Nb_{2/3}O_3$-8% $PbTiO_3$. Physics Review Letter 2001:86,3891-3895.

[28]   Shang JK. Tan X. Indentation-induced domain switching in PMN-PT piezoelectric crystal. Acta Materialia 2001:49,2993-2999.

[29]   Sani A. Noheda B. Kornev IA. High-pressure phases in highly piezoelectric $PbZr_{0.52}Ti_{0.48}O_3$. Physics Review B 2004:69,020105-020108.

[30]   Burcsu E. Ravichandran G. Bhattacharya K. Large strain electrostrictive actuation in barium titanate. Applied Physics Letter 2000:77,1698-1670.

[31]   Shu YC. Bhattacharya K. Domain patterns and macroscopic behaviour of ferroelectric materials. Philosophical Magazine Part B 2001:81(12),2021-2054.

[32]   Yu HF. Zeng HR. Wang HX. Li GR. et al. Domain structures in tetragonal $Pb(Mg_{1/3}Nb_{2/3})O_3$-$PbTiO_3$ single crystals studied by piezoresponse force microscopy. Solid State Communications 2005:133,311-314.

[33]   Ye ZG. Bing Y. Gao J. Bokov AA. Development of ferroelectric order in relaxor (1-x)$Pb(Mg_{1/3}Nb_{2/3})O_3$-x$PbTiO_3$(0<~x<~0.15). Physics Review B. 2003:67,104104-104107.

[34]   Ghosez P. Gonze X. Michenaud JP. First-principles characterization of the four phases of barium titanate. Ferroelectrics 1999:220,1-15.

[35]   Shuo D. Yong Z. Long LY. Elastic and Piezoelectric Properties of $Ce:BaTiO_3$ Single Crystals.Chinese Physics Letter 2005:22(7),1790-1792.

[36]   Feng ZY. Tan OK. Zhu WG. et al. Aging-induced giant recoverable electrostrain in Fe-doped $0.62Pb(Mg_{1/3}Nb_{2/3})O_3$-$0.38PbTiO_3$ single crystals. Applied Physics Letters 2008:92, 142910-142914.

[37] Weaver PM. Cain MG. Stewart M. Room temperature synthesis and one-dimensional self-assembly of interlaced Ni nanodiscs under magnetic field. J. Phys. D: Appl. Phys. 2010:43,275002-275006.

[38] Desheng F. Takahiro A. Hiroki T. Ferroelectricity and electromechanical coupling in (1− x)AgNbO$_3$-xNaNbO$_3$ solid   solutions.   Applied   Physics   Letter 2011:99,012904-012907.

Phase Transitions, Dielectric and Ferroelectric Properties of Lead-free NBT-BT Thin Films

N. D. Scarisoreanu, R. Birjega, A. Andrei, M. Dinescu,
F. Craciun and C. Galassi

Additional information is available at the end of the chapter

1. Introduction

Ferroelectric perovskites based on $Na_{0.5}Bi_{0.5}TiO_3$ (NBT) are considered among the most promising lead-free candidate materials to substitute $Pb(Zr_{1-x}Ti_x)O_3$ (PZT) in devices designed to respect standards and environmental laws. Taking into account the toxicity of lead-based systems, there are numerous lead-free piezoelectric materials under investigation in worldwide spread laboratories for replacing PZT in future devices. Constant efforts are made to find viable replacements for all these materials containing harmful elements.

Solid-solution systems based on lead-free perovskites like $Na_{0.5}K_{0.5}NbO_3$ (NKN), $BaTiO_3$ (BT), $Na_{0.5}Bi_{0.5}TiO_3$ (NBT) or bismuth layered-structured $SrBi_2Ta_2O_9$ (SBT), and $SrBi_2Nb_2O_9$ (SBN) are considered as viable alternatives for replacing lead-based materials. For example, $(K,Na)NbO_3$–$LiTaO_3$–$LiSbO_3$ alkaline niobate ceramics exhibit a d_{33} piezoelectric coefficient up to 416 pC/N together with Curie temperature T_c around 526 K, as reported by Saito et al [1]. Sodium/bismuth titanate (NBT) belongs to the bismuth-based perovskites in which the A-site atom is replaced. The crystalline structure, phase transitions and physical properties have been intensively studied since the discovery of the material in 1960 by Smolensky et al [2]. NBT has a relatively high depolarization temperature, T_d = 470 K, high remanent polarization, 38 $\mu C/cm^2$ and piezoelectric coefficient d_{33} = 125 pC/N [3]. However, owing to the high value of the coercive field and high electrical conductivity, NBT cannot be easily polarized, therefore different A-site substitutions have been attempted to avoid this drawback.

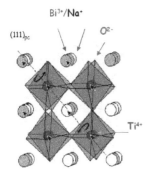
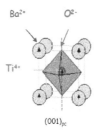

Figure 1. The end-members of perovskite NBT-BT: rhombohedral NBT and tetragonal BT. Cations Na^+/Bi^{3+} and Ba^{2+} occupy the A-sites while Ti^{4+} occupies B-sites (oxygen octahedra centers).

The solid solution with $BaTiO_3$, (1-x) NBT-x BT shows a morphotropic phase boundary (MPB) between the rhombohedral and the tetragonal phase, at x between 0.06 and 0.07 for which the material properties are considerably improved. Indeed d_{33} values up to 450 pC/N, and huge electric field-induced strain have been reported [4, 5]. Figure 1 shows the crystalline structures of the NBT and BT end members at room temperature. Perovskite structure deformations include oxygen octahedral rotations around different axis and cation shifts, therefore giving rise to a complex succession of ferroelastic and ferroelectric phase transformations with temperature variation.

Due to this polymorphic structure, NBT and NBT-BT have been also intensively studied in order to clarify their complicated phase transitions, which still pose questions [6].Structural and polar transformations in NBT-BT are more complicated than in other perovskite solid solutions, also due to the strong disorder of the A-sites occupied by Na^+, Bi^{3+} or Ba^{2+} ions, with different valence, mass and ionic radius. NBT transforms successively, from the high temperature cubic paraelectric into tetragonal antiferroelectric (or ferrielectric) and further into a rhombohedral ferroelectric phase [6]. In solid solution with BT, the ground ferroelectric phase changes from rhombohedral $R3c$ to tetragonal ferroelectric $P4mm$, at the so-called morphotropic phase boundary (MPB) ($x \approx 0.06\text{-}0.07$) [5, 7, 8]. The phase diagram of NBT-BT bulk material, mainly based on dielectric measurements, was completed by Cordero *et al* by performing direct anelastic measurements, the border between tetragonal and cubic phases being evidenced [9, 10, 11].

For NBT-BT thin films growth many techniques have been used. Guo *et al.* have investigated NBT-BT-based tri-layered films prepared by chemical solution deposition as a possible solution to the problem of avoiding leakage currents under high electric fields [12, 13]. Using pulsed laser deposition (PLD), Duclère *et al.* have reported the heteroepitaxial growth of NBT thin films on epitaxial platinum electrodes supported on a sapphire substrate [14]. More recently, M.Bousquet *et al* have described the electrical properties of (110)- oriented NBT thin films deposited by laser ablation on (110)Pt/(110) SrTiO$_3$ substrates [15]. They reported the coexistence of two kinds of grains with different shapes in the films, flat and elongated grains corresponding to (100) and (110) oriented NBT crystallites. The effects of Bi- excess in target on the dielectric and ferroelectric properties of the films have been also presented; the reported values for relative permittivity and remnant polarization were $\varepsilon_r \approx 225\text{-}410$ and 14 μC/cm^2, respectively. Furthermore, very recently, the electrical properties of (100)-oriented Na$_{0.5}$Bi$_{0.5}$TiO$_3$-BiFeO$_3$ thin films deposited by sol-gel have been reported by Qin et al, aiming to important applications such as photovoltaic devices [16].

However, despite the fact that ferroelectric materials with MPB have enhanced ferroelectric and piezoelectric properties, it is difficult to transpose them in thin films since MPB is limited to a small composition range. Almost all the physical parameters involved in thin films deposition like the substrate type, the microstructure and stress have strong impact on their physical properties [17]. In some previous papers we have investigated the role of different deposition parameters on NBT-BT film growth and properties [18, 19]. In this chapter, we discuss the role of certain experimental conditions like deposition temperature and substrate type, as well as of the amount of BT present in the target on crystalline structure, microstructure, dielectric properties, phase transition temperatures and stability limits of ferroelectric phases in NBT-BT thin films produced by PLD.

2. Experimental method

Pulsed laser deposition (PLD) was used for the film growth. The targets with composition (NBT)$_{1-x}$(BT)$_x$ (x = 0.06-0.08), further called NBT-BT6 and NBT-BT8, have been prepared following the mixed oxide route and sintered at 1150 °C for 2 h. The sintering was performed in crucibles with the sample surrounded with NBT pack, in order to avoid the loss of Na and Bi, which occurs at temperatures over 1000 °C; more details can be found in Ref.11. X-ray diffraction analysis evidenced the obtaining of pure perovskite phase. The microstructure of the sintered targets was investigated on polished and etched surfaces by scanning electron microscopy. The observed grain sizes were 2-10 μm.

For the film deposition, a Surelite II Nd:YAG pulsed laser with wavelength of 265 nm, pulse duration of 5 ns and frequency 10 Hz, has been employed. The laser fluence was set at 1.6 J/cm^2. The films were grown on Nb:STO and Pt/TiO$_2$/SiO$_2$/Si substrates, placed at a distance of about 4.3 cm from the target. Different sets of films have been grown at different substrate temperatures, ranging between 650-730 °C. Deposition and after-deposition cooling were performed in flowing oxygen atmosphere (0.3-0.6 mbar) to favour the formation of perovskite phase without oxygen vacancies. Chemical composition was checked via SIMS technique using a Hiden SIMS/SNMS system. The thickness of the thin films, evaluated by spectroellipsometry, was between 300-500 nm.

For the investigation of the crystalline structure of the targets and films, a PANalytical X'pert MRD diffractometer in Bragg-Brentano geometry was used. The measurements were performed with a step size of 0.02^0 and with a scanning time on step of 25 s or 250 s, depending on the angular range.

The film surface morphology was examined by AFM (model XE100, Park Systems). Piezoelectric force microscopy measurements were performed with a PFM system which includes a lock-in amplifier SR-830 and a dc- high voltage amplifier WMA-280. Conductive all-metal Pt tips were employed for these measurements were the switching characteristcs of the films have been tested.

Several Au electrode dots with an area of about 0.22 mm^2 have been evaporated through a mask on the films for electrical characterization. Polarization hysteresis was measured by using a Radiant Technology RT66A ferroelectric test system, in the virtual ground mode. The dielectric measurements were carried out in a frequency range between 200 Hz and 1 MHz using an HP 4194A impedance analyzer and an HP 4284A LCR meter with a four wire probe. The measurements were performed at 1.5 K/min between 300 and 570 K in a Delta Design climatic chamber model 9023 A (on targets) and in a Linkam variable-temperature stage (model HFS 600E) on films.

3. Results and discussion.

3.1. Growth mode of NBT-BT thin films.

The microstructure of ceramic thin films is one of the most important factors that influence their physical properties. Since the growth mode of thin films is strongly dependent on the substrate type, we investigated the deposition of NBT-BT thin films on two different types of substrates:

1.   single crystal $SrTiO_3$: Nb (Nb:STO) and

2. $Pt/TiO_2/SiO_2/Si$.

The AFM pictures obtained on the two sets of films show important microstructural differences, mainly due to different growth mechanisms. In Figure 2 we show AFM images taken on a NBT-BT6 film deposited on Nb: STO monocrystalline substrate at 650 ^{0}C. It can be observed that a first stage of growth resulting into a continuous layer stops when the critical thickness for misfit dislocations (probably a few tens of nm) is reached. After that, the growth continues in platelet-like form (see details in Fig. 2 b). If the deposition temperature is not sufficiently high to favor material exchange between platelets via surface migration, successive layers will grow on the top of the first islands and the growth will result into a discontinuous layer. This explains the platelet-like aspect of the film shown in Figure 2.

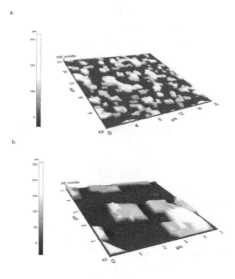

Figure 2. AFM images of NBT-BT6% films deposited on Nb:STO substrates at temperature of 650 °C. The displayed surfaces are 20x20 µm² (a) and 5x5 µm² (b).

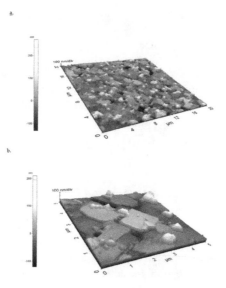

Figure 3. AFM images of NBT-BT6% films deposited on Nb:STO substrates at 700°C. The displayed surfaces are 20x20 µm² (a) and 5x5 µm² (b).

However, raising the substrate temperature to 700 °C during the deposition of a second set of films while keeping constant all the other parameters, including the number of laser pulses, produces a uniform layer of continuous platelets, on top of which new islands nucleate (Figure. 3).

A rather different morphology is displayed by NBT-BT films grown on Pt/TiO$_2$/SiO$_2$/Si (Fig. 4 and 5). In this case, the growth progresses from the beginning in island-like form since the polycrystalline Pt layer provide the nucleation sites for their formation. Moreover, these NBT-BT islands grow on the Pt layer without preserving a unique orientation, due to the same reason. Instead films grown on Nb:STO monocrystalline structures are uniaxially (001)-oriented, as it will be shown in the next section.

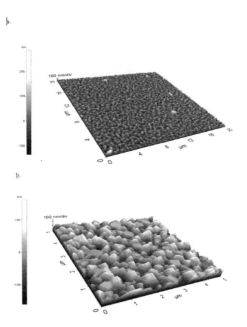

Figure 4. AFM images of NBT-BT6% films deposited on Pt/TiO$_2$/SiO$_2$/Si substrate at a temperature of 700 °C. The displayed surfaces are 20x20 µm² (a) and 5x5 µm² (b).

A fine microstructure with grain size ranging from a few tens of nm up to a few hundred of nm is displayed by NBT-BT6 films (Fig. 4). We note the striking difference with bulk samples microstructures (not shown here), which consists of crystallites of 1-10 µm size.

A similar fine microstructure is displayed by NBT-BT8% films grown on Pt/TiO$_2$/SiO$_2$/Si (Fig. 5 a). However, the enlarged AFM image displayed in Fig 5 b) reveals a somewhat different aspect with triangular nanograins lying in plane.

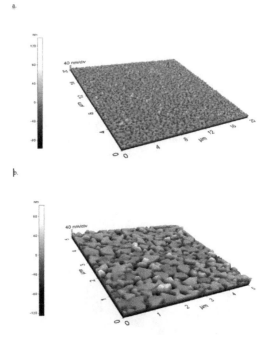

Figure 5. AFM images of NBT-BT8% films deposited on Pt/TiO$_2$/SiO$_2$/Si substrate at a temperature of 700 ° C. The displayed surfaces are 20x20 μm^2 (a) and 5x5 μm^2 (b).

3.2. Crystalline structure

The XRD spectrum of NBT-BT6 target corresponds to a mixture of rhombohedral R3c and tetragonal P4mm phases, as shown by the splitting of (111) and (200), (012) and (024) rhombohedral peaks in the bottom pattern in Fig. 6 [20]. The main Miller index of the rhombohedral phase are depicted horizontally on the bottom of the figures while those of the tetragonal phase vertically above. On the same graph, the pattern corresponding to the NBT-BT6 film grown on Pt/TiO$_2$/SiO$_2$/Si at 700 °C is given. The curve corresponding to NBT-BT6/Pt/TiO$_2$/SiO$_2$/Si film deposited in the same conditions but at a substrate temperature of 650^0 exhibits similar features as we had reported and is not presented here [18].The as deposited thin films exhibit pure perovskite phase with symmetry congruent with that of the target. The reflection peaks indicate a randomly oriented structure, consistent with the polycrystalline nature of the films.

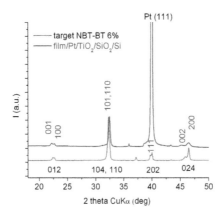

Figure 6. XRD spectra of NBT-BT6% deposited on Pt/TiO$_2$/SiO$_2$/Si substrate. The bottom pattern corresponds to the target

Figure 7 displays the XRD patterns of NBT-BT6 films deposited at two temperatures, 650 °C and 700 °C, around the (100)/(001) and (200)/(002) reflections of the Nb:STO substrate. The spectra indicate the epitaxial growth of NBT-BT6% films on the Nb:STO substrate at the two temperatures. This feature is congruent with the microstructure shown in the previous section (Fig. 2 and Fig. 3), consisting of large platelet-like crystallites which preserve the same axis of orientation with the monocrystalline substrate.

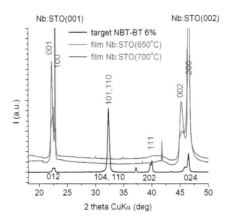

Figure 7. XRD spectra of NBT-BT6% /Nb:STONb:STO films deposited at 650 °C and 700 °C. The grey pattern represents the Nb:STONb:STO target reflection peaks.

Figure 8 shows the XRD patterns of NBT-BT8 films grown on Pt/TiO$_2$/SiO$_2$/Si and Nb:STO substrates at 700 °C. The grey pattern represents the NBT-BT8 target spectrum, which corresponds to the tetragonal P4mm symmetry. It can be observed that, similar to the previous composition, the growth on single crystal Nb:STO substrate produces an epitaxial film, while the growth on Pt/TiO$_2$/SiO$_2$/Si substrate results into a polycrystalline randomly oriented film.

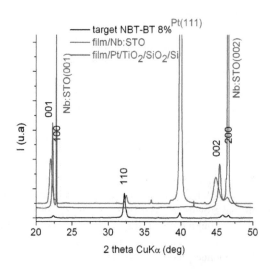

Figure 8. XRD spectra of NBT-BT8% deposited on Pt/TiO$_2$/SiO$_2$/Si and on Nb:STONb:STO substrates

3.3. Dielectric and ferroelectric properties

The dielectric and ferroelectric properties of NBT-BT thin films have been evaluated on capacitors formed by evaporating through a mask an array of gold electrode dots with an area of about 0.22 mm^2 on the surface of films grown on Pt/TiO$_2$/SiO$_2$/Si and Nb:STO substrates. The bottom electrode was formed by the Pt layer in the first case or by the Nb:STO substrate itself in the second case.

The piezoresponce force microscopy results are presented in Figure 9. The full-Pt tips were brought in contact with the surface of the sample and then a *dc* bias and test *ac* bias were applied between the tip and the bottom electrode of the samples. The *dc* bias was generated by a high voltage amplifier and the *ac* bias was generated by a lock-in amplifier. The same lock-in amplifier was used to analyse the vertical deflection signal from the PSPD, in order to extract the amplitude and the phase of the cantilever oscillations induced by the local deformation of the sample due to the applied *dc* bias. The NBT-BT6/Pt/Si thin films show good switching behavior, the piezoelectric hysteresis and pronounced imprint (not showed here) confirming the piezoelectric and ferroelectric characteristics. The dependence of effective piezoelectric coefficient d_{33} ^{eff} on the applied electric field is given in Figure 9. The locally measured values with the highest being around d_{33} $^{eff} \approx$ 83 pm/V, are even higher then for previouslly reported values for pure NBT or lead-based thin films, such as Pb(ZrTi)O$_3$ or PbTiO$_3$ [21, 22]. However, these d_{33} ^{eff} are a bit smaller then NBT-BT6 ceramics which are reported to be more than 100 pm/V [21]. The reasons for these smaller values are related with the film's porosity, but also with the clamping effect which occurs because the PFM tip- applied electric field will piezoelectrically deform only a small fraction of the film. The rest of the sample will restrict the relative deformation of this small fraction, resulting a lower value for d_{33} ^{eff} [5, 22].

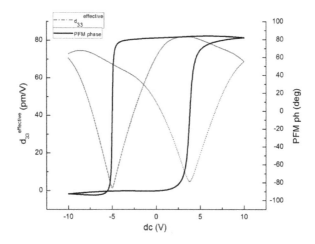

Figure 9. The piezoresponse measurements performed on NBT-BT6 thin films.

In Fig. 10 the room temperature dielectric properties of NBT-BT6 films deposited on Pt/ TiO$_2$/SiO$_2$/Si at different substrate temperatures, 650 °C and 730 °C, have been compared in the frequency range 100 Hz-1 MHz. Films grown at 650 °C show a higher dielectric constant (ε' ~ 1000), in the order of magnitude of the bulk values (ε'_{bulk} ~ 1900), while films grown at 730 °C show lower values (ε' ~ 700). The dielectric loss values are instead comparable in the two samples, and similar to bulk values.

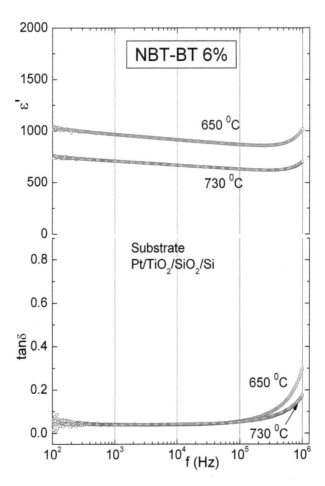

Figure 10. Room temperature dielectric constant ε' and loss tanδ variation with frequency for NBT-BT6% films deposited at different temperatures on Pt/TiO₂/SiO₂/Si.

Figure 11 displays the room temperature dielectric constant and dielectric loss in the frequency range 100 Hz-1 MHz for NBT-BT8 films grown on Pt/TiO₂/SiO₂/Si at different temperatures: 650, 700 and 730 °C. Unlike the previous composition, in this case growth at higher substrate temperatures was beneficial for the improvement of dielectric properties, at least in the frequency domain up to a few hundred kHz. Above this frequency there is a strong increase of dielectric loss. Since an increase is registered also in the dielectric constant, this could be caused by a relaxation mechanism which is active at room temperature at these frequencies, like e.g. free charge relaxation.

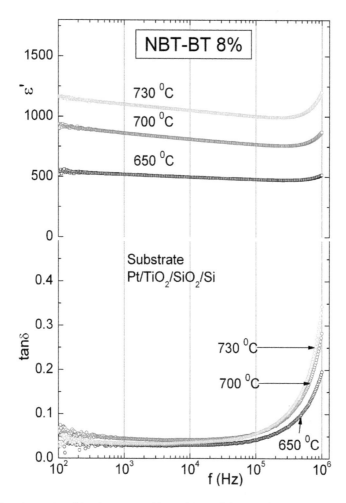

Figure 11. Room temperature dielectric constant and loss variation with frequency for NBT-BT8% films deposited at different temperatures on Pt/TiO₂/SiO₂/Si

Polarization hysteresis measurements on NBT-BT6 films grown on Pt/TiO$_2$/SiO$_2$/Si are shown in Fig. 12. Spontaneous polarization was about 30 μC/cm^2 and the remnant polarization was about 10 μC/cm^2. The rather high value of coercive field (100 kV/cm) could be explained by the presence of intrinsic strain and pinning defects.

Dielectric and ferroelectric properties measurements on films deposited on Nb:STO substrates have been less reliable, probably due to the presence of a non-ohmic contact at the NBT-BT film – semiconductor Nb:STO interface. However PFM measurements (not shown here) evidenced good piezoelectric response, which indicates good intrinsic dielectric and ferroelectric properties, although quantitative values are difficult to extract.

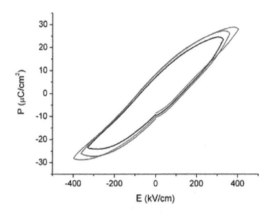

Figure 12. Polarization-electric field hysteresis loop measured on a NBT-BT6% film deposited on Pt/TiO$_2$/SiO$_2$/Si

3.4. Phase transitions

Phase transitions in ferroelectric materials are accompanied by anomalies of complex dielectric permittivity variation with temperature, generally narrow peaks or steps, depending on the type of phase transformation. However NBT-BT compositions near the morphotropic phase boundary behave as relaxors, due to the cation disorder. This is evidenced in Fig. 13 for NBT-BT8 bulk material. The main characteristic of a relaxor ferroelectric is a broad dielectric peak at a temperature T_m which is not related to a structural transformation. This is due to a wide distribution of relaxation times which characterizes the dielectric response of polar nanoregions. This peak shifts with the increasing of the measurement frequency toward higher temperatures. Thus the dielectric maximum of NBT-BT8 shifts from about 497 K at 200 Hz to about 512 K at 100 kHz. A similar dependence is obeyed also by the dielectric loss.

In Fig. 14 the variation of dielectric constant and loss with temperature for a NBT-BT8 film deposited on Pt/TiO$_2$/SiO$_2$/Si is shown. The maximum of the dielectric constant occurs at about 485 K, not far from bulk T_m. However the anomaly is characteristic of a well-behaved phase transition, since the peak temperature T_m does not shift with frequency. A similar qualitative behavior was observed also on the dielectric permittivity variation with temperature for NBT-BT6 films (not shown here).

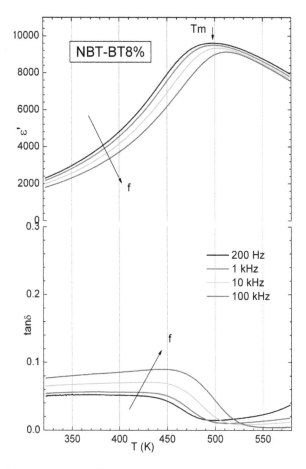

Figure 13. The dielectric permittivity and loss variation with temperature measured on a NBT-BT8% bulk sample at different frequencies. The long arrows mark the increasing of frequency.

While the dielectric permittivity peak position Tm ~ 485 K does not shift with measuring frequency, peak height is strongly dependent on it, decreasing for higher frequencies. This can be attributed to the possible presence of a non-polar dielectric layer, which does not influence the general behavior at phase transition, but can modify the value of the dielectric constant [23]. Generally these interface layers can have strong frequency-dependent dielectric properties which influence the overall properties of the heterostructures. However we stress again that the dependence on temperature and phase transition temperatures can be influenced only in the limits of a monotonous contribution, since the dielectric behaviour of non-polar layers is free of temperature anomalies. Indeed the stronger variation with temperature of the peak intensity at T_m could be attributed to the non-polar layer contribution at higher temperatures due to conductivity variation.

The temperature Td where a strong increase of dielectric loss and dielectric constant occurs marks the ferroelectric – antiferroelectric phase transition which, in bulk samples with the same composition is visible only in the poled state. It is called also depolarization temperature.

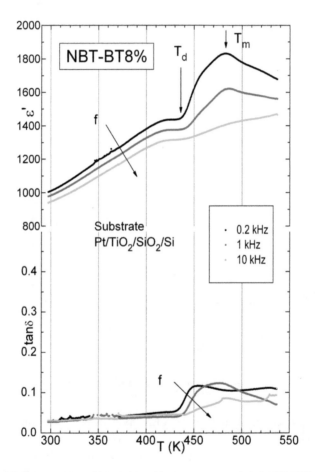

Figure 14. The dielectric permittivity and loss variation with temperature measured on a NBT-BT8% thin film at different frequencies. The long arrows mark the increasing of frequency.

An apparent frequency dependence of the step increase in tanδ which marks T_d is visible on the lower curves in Fig. 14. This could be attributed to a partial relaxor behavior due to mixed nanodomains-normal ferroelectric domains, which can be find also, but in proportion displaced to the nanodomain limit, in bulk samples. We must remark that the films were not poled, since no bias electric field was applied. Therefore the occurrence of a ferroelectric ground state in the NBT-BT films, in striking contrast with relaxor bulk samples with the same composition, must be generated by some intrinsic differences between ceramic bulk samples and ceramic thin films. The first and most obvious reason could be related to the constraining stress of the substrate on the thin films. This is strong in epitaxially grown thin films with a thickness generally below 100 nm, but it should be almost absent in polycrystalline films, randomly oriented and with a thickness of several hundreds of nm. This last one is the case for NBT-BT thin films deposited on $Pt/TiO_2/SiO_2/Si$ substrates. The second reason could be related to the strong differences in the microstructures of ceramic thin films and bulk materials with the same composition. Therefore the occurrence of a ferroelectric ground state instead of a relaxor state in NBT-BT films, as well as the occurrence of a true ferroelectric phase transition could be due to the constrainig imposed by the nanograin boundaries on the ensemble of polar nanoregions.

4. Conclusions

In summary, we have investigated the role of deposition temperature and substrate type as well as the amount of BT present in the target on crystalline structure, microstructure, dielectric properties, phase transition temperatures and stability limits of ferroelectric phases in NBT-BT thin films grown by pulsed laser deposition. We have successfully deposited pure perovskite epitaxial films on single-crystal Nb:STO substrates. Successful growth of NBT-BT films on platinized silicon substrates has been achieved. Good dielectric and ferroelectric properties, comparable with bulk values, have been obtained. The NBT-BT6/Pt/Si thin films show a classic switching behavior, the piezoelectric hysteresis and pronounced imprint confirming the piezoelectric and ferroelectric characteristics. The locally measured value of effective piezoeletric coefficient d_{33}^{eff} was around 83 pm/V, higher to the previouslly reported values for pure NBT or lead-based thin films. An enhanced stability of ferroelectric phase in thin films with respect to bulk has been observed and explained by their peculiar nanocrystalline microstructure.

Author details

N. D. Scarisoreanu[1*], R. Birjega[1], A. Andrei[1], M. Dinescu[1], F. Craciun[2] and C. Galassi[3]

*Address all correspondence to: snae@nipne.ro

1 NILPRP, National Institute for Laser, Plasma & Radiation Physics, Bucharest, Romania

2 CNR-ISC, Istituto dei Sistemi Complessi, Area della Ricerca Roma-Tor Vergata, Roma, Italy

3 CNR-ISTEC, Istituto di Scienza e Tecnologia dei Materiali Ceramici, Faenza, Italy

References

[1] Saito, Y., Takao, H., Tani, T., Nonoyama, T., Takatori, K., Homma, T., Nagaya, T., & Nakamura, M. (2004). Lead-free piezoceramics. *Nature*, 432, 84.

[2] Smolenski, G. A., Isupov, V. A., Agranovskaya, A. I., & Krainik, N. N. (1961). New ferroelectrics of complex composition, Sov. *Phys. Solid State*, 2, 2651-196.

[3] Chiang, Y. M., Farrey, G. W., & Soukhojak, A. N. (1998). Lead-free high-strain single-crystal piezoelectrics in the alkaline-bismuth-titanate perovskite family. *Appl. Phys. Lett*, 73, 3683.

[4] Takenaka, T., & Nagata, H. (1999). Present Status of Non-Lead-Based Piezoelectric-Ceramics. *Key Engineering Materials*, 157/158, 57.

[5] Takenaka, T, Maruyama, K.-I., & Sakata, K. (1991). (Bi1/2Na1/2)TiO3-BaTiO3 System for Lead-Free Piezoelectric Ceramics. *Jpn. J. Appl. Phys*, 30, 2236-2239.

[6] Jones, G. O., & Thomas, P. A. (2002). Investigation of the structure and phase transitions in the novel A-site substituted distorted perovskite compound Na0.5Bi0.5TiO3, Acta Crystallogr. *Sect. B: Struct. Sci*, 58, 168 -178 .

[7] Hiruma, Y., Watanabe, Y., Nagata, H., & Takenaka, T. (2007). Phase Transition Temperatures of Divalent and Trivalent Ions Substituted (Bi1/2Na1/2)TiO3 Ceramics, Key Eng. *Mater*, 350, 93.

[8] Glazer, A. M. (1972). The classification of tilted octahedra in perovskites, Acta Crystallogr. *Sect. B: Struct. Sci*, 28, 3384.

[9] W.-van, B., Eerd, D., Damjanovic, N., Klein, N., Setter, , & Trodhal, J. (2010). Structural complexity of (Na0.5Bi0.5)TiOP3BaTiO3 as revealed by Raman spectroscopy. *Phys. Rev. B*, 82, 104112.

[10] Ma, C., & Tan, X. (2010). Phase diagram of unpoled lead-free (1-x)(Bi/sub 1/2/Na/sub 1/2/)TiO/sub 3-x/BaTiO/sub 3/ ceramics. *Solid State Commun*, 150, 1497-1500.

[11] Cordero, F., Craciun, F., Trequattrini, F., Mercadelli, E., & Galassi, C. (2010). Phase transitions and phase diagram of the ferroelectric perovskite (Na0.5Bi0.5)1−xBaxTiO3 by anelastic and dielectric measurements. *Phys. Rev. B*, 81, 144124.

[12] Guo, Y. P., Akai, D. S., Sawada, K., & Ishida, M. (2008). Dielectric and ferroelectric properties of highly (100)-oriented (Na0.5Bi0.5)0.94Ba0.06TiO3 thin films grown on LaNiO3/γ-Al2O3/Si substrates by chemical solution deposition. *Solid State Sciences*, 10, 929.

[13] Guo, Y. P., Akai, D. S., Sawada, K., Ishida, M., & Gu, M. Y. (2009). Structure and electrical properties of trilayered BaTiO3/(Na0.5Bi0.5)TiO3BaTiO3/BaTiO3 thin films deposited on Si substrate. *Solid State Communications*, 149, 14.

[14] Duclère, J.R., Cibert, C., Boulle, A., Dorcet, V., Marchet, P., Champeaux, C., Catherinot, A., Députier, S., & Guilloux-Viry, M. (2008). Lead-free Na0.5Bi0.5TiO3 ferroelec-

tric thin films grown by Pulsed Laser Deposition on epitaxial platinum bottom electrodes. *Thin Solid Films*, 517, 592.

[15] Bousquet, M., , J., Ducle`re, R., Gautier, B., Boulle, A., Wu, A., De´, S., putier, D., Fasquelle, F., Re´mondie`, re. D., Albertini, C., Champeaux, P., Marchet, M., Guilloux- , Viry, & Vilarinho, P. (2012). Electrical properties of (110) epitaxial lead-free ferroelectric Na0.5Bi0.5TiO3 thin films grown by pulsed laser deposition: Macroscopic and nanoscale data. *J. Appl. Phys*, 111, 104 -106.

[16] Qin, W., Guo, Y., Guo, B., & Gu, M. (2012). Dielectric and optical properties of BiFeO3-(Na0.5Bi0.5)TiO3 thin films deposited on Si substrate using LaNiO3 as buffer layer for photovoltaic devices. *J. Alloys Compd*, 513, 154-158.

[17] Shaw, T. M., Trolier Mc Kinstry, S., & Mc Intyre, P. C. (2000). The properties of ferroelectric thin films at small dimensions. *Annu. Rev. Mater. Sci*, 30, 263.

[18] Scarisoreanu, N., Craciun, F., Chis, A., Birjega, R., Moldovan, A., Galassi, C., & Dinescu, M. (2010). Lead-free ferroelectric thin films obtained by pulsed laser deposition. *Appl. Phys A*, 101, 747 -751 .

[19] Scarisoreanu, N., Craciun, F., Ion, V., Birjega, S., & Dinescu, M. (2007). Structural and electrical characterization of lead-free ferroelectric Na/sub 1/2/Bi/sub 1/2/TiO/sub 3/-BaTiO/sub 3/ thin films obtained by PLD and RF-PLD. *Appl. Surf. Sci*, 254, 1292.

[20] Picht, G., Töpfer, J., & Henning, E. (2010). Structural properties of (Bi0.5Na0.5)1–xBaxTiO3 lead-free piezoelectric ceramics. *J. Eur. Ceram. Soc*, 30, 3445.

[21] Takenaka, T. (1999). Piezoelectric Properties of Some Lead-Free Ferroelectric Ceramics. *Ferroelectrics*, 230, 87-98.

[22] Morelli, A., Sriram, Venkatesan, Palasantzas, , Palasantzas, Kooi, G, & De Hosson, J. Th. M. (2009). Piezoelectric properties of PbTiO3 thin films characterized with piezoresponse force and high resolution transmission electron microscopy. *J. Appl. Phys*, 105, 064106.

[23] Craciun, F., & Dinescu, M. (2007). Piezoelectrics", in "Pulsed laser deposition of thin films: applications-led growth of functional materials". *edited by R. Eason (Wiley-Interscience, John Wiley & Sons, Hoboken, New Jersey, U.S.A.,)*, 487-532.

Chapter 4

'Universal' Synthesis of PZT (1-X)/X Submicrometric Structures Using Highly Stable Colloidal Dispersions: A Bottom-Up Approach

A. Suárez-Gómez, J.M. Saniger-Blesa and
F. Calderón-Piñar

Additional information is available at the end of the chapter

1. Introduction

The synthesis of nanostructured ferroelectric materials has been a subject of increasing interest for more than a decade due to their possible applications as nonvolatile/dynamic random access memories (NVRAMs/DRAMs), tunnel effect capacitors for high frequency microwave applications, infrared detectors, micro-electromechanical systems (MEMs) and electro-optical modulators among others. These materials, when fully integrated into appropriately designed micro-systems, are promising candidates for robotics sensing and for future medical procedures requiring an in situ, real time and nondestructive monitoring with very high sensitivity.

As a logical consequence, the quest for nanostructured ferroelectric materials has also led to modify the synthesis routes usually considered for obtaining conventional bulk materials and thin films. One of those routes, the sol–gel method, overcomes the inherent limitations of the conventional powders-based synthesis routes when dealing with molecular homogeneity and, therefore, it has been extensively studied and applied in different scenarios in order to obtain not only thin films, nanotubes or nanorods, but also submicron grains due to the distinctive dielectric and ferro/piezoelectric features that are associated with size related effects when average grain size is well below 1 μm [1]-[5]and because of the potential use of these materials in a plethora of nanodevices [6].

One of the backbones of the ferroelectrics industry is the Lead zirconate titanate [PZT: $Pb(Zr_{1-x}Ti_x)O_3$] ceramic system, a well-known ABO_3 perovskite with a wide range of industrial applications since the early 1950s of the 20th century. Its dielectric and electromechanical properties have made possible to develop and implement a myriad of devices such as underwater sonar systems, sensors, actuators, accelerometers, ultrasonic equipment, imaging devices, microphones and multiple active and passive damping systems for the car industry.

The sol–gel synthesis of PZT ceramics has evolved a lot since it was first reported in the mid-1980s of the last century [7]-[9]; thermal treatments, precursors, additives, diluents and stabilizers have been incorporated and/or modified in order to achieve less hydrolysable, more stable, compounds to suit a specific need. In the case of the synthesis of thin films, nanotubes or nanorods, sol–gel routes have been used in many different ways: from the well-known spin-coating method to the electrophoretic deposition on a given substrate [10] or the insertion on nanoporous templates [11]. Submicron and nanosized PZT particles are also important in the fabrication of highly dense bulk ceramics which, even nowadays, comprise almost the entirety of the electroceramics market. In this particular case, several works have been published in which PZT powders are synthesized first via sol–gel and then put to sinter for densification [12]-[15]. In this procedure, sintering temperatures tend to lower and densification attains notably high values mostly due to a higher Gibb's free energy per unit surface area while some material properties strongly dependent on grain size are also enhanced [16].

Accordingly, in this chapter we will be devoted to analyze the feasibility of a 'customized' synthesis of $PbZr_{1-x}Ti_xO_3$ [PZT (1-x)/x] nano/submicrometric structures by using a sol-gel based colloidal dispersion as a precursor solution. This study will be done on the basis of a 'bottom-up' approach as it will take into account (i) Synthesis route, (ii) Properties of colloidal dispersions and (iii) Final crystallization. We will try to thoroughly illustrate every synthesis step while paying special attention to the physicochemical depiction of some phenomena that not always are sufficiently described or explained. It has to be pointed out that most of the main procedures, discussions and results contained herein could be easily extrapolated to a wide range of materials, not exclusively PZT-based ones.

2. Sol-gel based synthesis route

The complexity of the intermediate reactions, one of the few handicaps of the sol–gel method, makes almost mandatory a step by step study of the synthesis method. The chemical reactivity of precursors is a well-known key feature determining the nature of the intermediate organic ligands and the control of the hydrolysis rate in the final sol–gel. Several studies have been made for particular synthesis routes both theoretical and experimental and many possible reaction pathways have been proposed and dissected. Particularly, the -chemistry of metal alkoxides has been intensively studied by Sanchez et al. [17] and Sayer and coworker [18]. It is our purpose here to analyze the different reaction steps and intermediates involved in the PZT (1-x)/x sol–gel synthesis using propoxides as starting metal alkoxides.

2.1. Synthesis

The followed sol-gel route was the acetic acid, acetylacetone and 2-methoxyethanol propoxy-based sol-gel method as illustrated in Figure 1. Starting reagents for the sol-gel PZT (1-

x)/x solutions were: (1) lead(II) acetate trihydrate (Pb(OAc)$_2$•3H$_2$O, Mallinckrodt Baker, Inc., 99.8% pure), (2) glacial acetic acid (HOAc, Mallinckrodt Baker, Inc., 99.7% pure), (3) acetyla-cetone (AcacH, Sigma-Aldrich Co., 99% pure), (4) 2-methoxyethanol (2-MOE, Mallinckrodt Baker, Inc., 100% pure), (5) zirconium(IV) propoxide (Zr(OPr)$_4$, Sigma-Aldrich Co., 70 wt.% in 1-propanol), and (6) titanium(IV) propoxide (Ti(O^{i}Pr)$_4$, Sigma-Aldrich Co., 97% pure).

First, lead acetate was dissolved in acetic acid with a 1:3 molar ratio while stirred and re-fluxed at 115 °C during 3 h for dehydration and homogeneity purposes. After this step, a thick transparent solution was obtained which will be referred hereafter as solution A. On a separate process, stoichiometric amounts of zirconium and titanium propoxides were mixed with acetylacetone on a 1:2 molar ratio in order to avoid fast hydrolysis of reactants. This mixture was stirred and refluxed at 90 °C during 4 h forming a clear yellow solution refer-red hereafter as solution B. When this solution cooled down, the precipitation of several nee-dle shaped crystals was verified. In order to keep stoichiometry unaffected, we then repeated the solution B procedure to isolate and characterize some of those crystals by sin-gle crystal XRD.

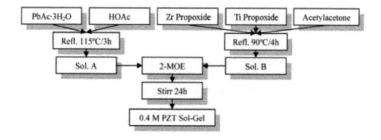

Figure 1. Flow chart depicting the basic experimental procedure followed in order to obtain a 0.4M PZT (1-x)/x sol–gel based precursor.

Solutions A and B were then mixed together as appropriate amounts of solvent (2-MOE) were slowly added for complete dilution of the precipitated crystals and for controlling the PZT concentration on the final solution. Afterwards, a light yellow solution was obtained af-ter stirring for 24 h at room temperature.

It must be stated here that we devoted our work to several Zr:Ti molar ratios that are com-monly used in practical applications: *(i)* PZT 25/75, *(ii)* PZT 53/47, *(iii)* PZT 60/40, *(iv)* PZT 80/20 and *(v)* PZT 95/05. Besides, we also tried to cover a relatively wide concentration range for every PZT (1-x)/x sol-gel precursor under study: from 0.05 M to 0.4 M. Due to these facts, we will only focus on the discussion of representative samples instead of discussing irrele-vant data of samples that showed no significant discrepancies between each other. As de-picted in Figure 1, this section will analyze the synthesis of a 0.4 M PZT 53/47 precursor solution.

2.2. Experimental techniques

Every reactant and intermediate product was analyzed using FT-IR and Raman spectroscopies. As was described earlier, single crystal XRD was also carried out for the solution B acicular precipitates.

2.2.1. Raman spectroscopy

Raman characterization was made on an Almega XR Dispersive Raman spectrometer equipped with an Olympus microscope (BX51). An Olympus 10x objective (N.A. = 0.25) was used both for focusing the laser on the sample, with a spot size ~ 5 μm, and collecting the scattered light in a 180° backscattering configuration. The scattered light was detected by a CCD detector, thermoelectrically cooled to -50 °C. The spectrometer used a grating (675 lines/mm) to resolve the scattered radiation and a notch filter to block the Rayleigh light. The pinhole of the monochromator was set at 25μm. The Raman spectra were accumulated over 25 s with a resolution of ~ 4 cm^{-1} in the 100-2000 cm^{-1} interval. The excitation source was a 532 nm radiation from a Nd:YVO$_4$ laser (frequency-doubled) and the incident power at the sample was of ~ 8 mW.

2.2.2. FT-IR spectroscopy

FT-IR analysis was carried out on a Thermo Nicolet Nexus 670 FT-IR in transmission mode with a resolution of ~ 4 cm^{-1} in the 400-2000 cm^{-1} interval. The excitation source was a Helium-Neon laser light incident on a KBr compact target containing ~ 0.5 % in weight of the sample of interest.

2.2.3. Single crystal XRD

Single crystal XRD experiments were carried out for selected specimens. A Bruker SMART APEX CCD-based X-ray three circle diffractometer was employed for crystal screening, unit cell determination and data collection. A Van Guard 40x microscope was used to identify suitable samples and the goniometer was controlled using the SMART software suite. The X-ray radiation employed was generated from a Mo sealed X-ray tube (Kα= 0.70173 Å with a potential of 50 kV and a current of 30 mA) and filtered with a graphite monochromator in the parallel mode (175 mm collimator with 0.5mm pinholes).

2.3. Solution A

Figure 2 shows, from bottom to top, the IR and Raman spectra of starting acetic acid and lead acetate as well as the final product, solution A, after stirring and refluxing. There are several features in common between these spectra due to the organic nature of ligands. Basically, bands corresponding to the CH$_2$ and CH$_3$ groups in the 1300-1400 cm^{-1} interval as well as in the low frequency range [19].

The glacial acetic acid spectra reveal several representative peaks found at 619 cm^{-1} (RA: Raman active), 889-893 cm^{-1} (RA, IRA: Infrared active), 1294 cm^{-1} (IRA), 1410 cm^{-1} (IRA), 1668 cm^{-1} (RA), 1716 cm^{-1} (IRA) and 1757 cm^{-1} (IRA) for this molecule. Those peaks correspond, respectively, to the τ(C-C=O), ν(C-C), δ_S(CH$_3$), δ_A(CH$_3$), ν(C=O), ν(C=O)$_{dimer}$ and ν(C=O)$_{mono-mer}$ vibrations [20],[21]. The clear splitting detected for the C=O dimeric and monomeric stretch vibrations somewhat evidences the presence of some small amounts of water in the acidic medium that, for the purpose of our study, will not be taken into consideration.

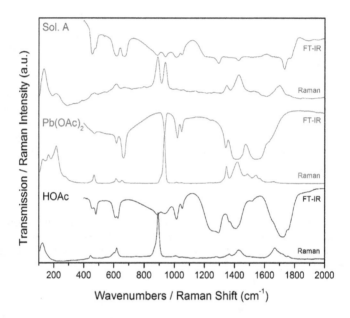

Figure 2. Infrared and Raman spectra of the reactants involved in solution A formation.

On the other hand, the lead(II) acetate trihydrate spectra also feature some representative peaks located at 216 cm^{-1} (RA), 615-617 cm^{-1} (RA, IRA), 665 cm^{-1} (IRA), 934-935 cm^{-1} (RA, IRA), 1342-1346 cm^{-1} (RA, IRA), 1417-1420 cm^{-1} (RA, IRA) and 1541-1543 cm^{-1} (RA, IRA) related to the ν(Pb-O), ϱ(COO), δ_S(COO), ν_S(C-C), δ_S(CH$_3$), ν_S(C-O) and ν_A(C=O) vibrational modes, respectively [20],[22]. It is well known that the acetate ligands can complex a metal ion in three different ways (monodentate, bidentate chelating and bridged) and that, unfortunately, none of these can be uniquely identified by symmetry considerations. Traditionally, the various types of bonding have been identified by the magnitude of the difference between symmetric ν_S(C-O) and asymmetric ν_A(C=O) vibrations. In our case, this difference is 122 cm^{-1} indicating a bidentate chelating coordination for the acetate-metal complex that has been widely accepted for lead(II) acetate even though the $\Delta\nu$ criterion has led sometimes to incorrect conclusions [20],[22].

Solution A vibrational spectra shown in Figure 2 evidences the expected dilution of lead(II) acetate in acetic acid: there are no new vibrational modes and a strong band overlapping is seen. There is some band shift due to the overlapping and is worth to notice the weakening of the lead acetate Raman active Pb-O band at 216 cm^{-1} when in solution. Unfortunately, a noticeable fluorescence in the Raman spectrum could not be avoided and this fact made very difficult to carry on an appropriate analysis with useful data.

Nevertheless, and according to our results, the formation of Solution A could be fairly described by the following equation:

$$Pb(OAc)_2 \cdot 3H_2O_{(s)} + 3HOAc_{(Aq)} \xrightarrow[1-3hrs.]{T-115^\circ C} Pb^{2+}_{(Aq)} + 3\alpha H_3O^+_{(Aq)} + (2+3\alpha)OAc^-_{(Aq)} +$$
$$+3(1-\alpha)HOAc_{(Aq)} \tag{1}$$

Where α is the HOAc dissociation degree under our experimental conditions and where we have also assumed, on the simplest approach, neglectable losses due to evaporation of H_2O and HOAc during the whole process.

2.4. Solution B

FT-IR spectra for Solution B precursors, Figure 3, showed some features like the ones discussed above. A Raman spectra based analysis for these compounds could not be completed due to the strong fluorescence exhibited by the zirconium and titanium alkoxides at our fixed operating laser wavelength.

The titanium propoxide IR spectrum exhibit sharp bands at 1377-1464 cm^{-1} corresponding, respectively, to the stretching and bending vibrations of the aliphatic CH$_3$ groups. A distinctive single peak at 1011 cm^{-1} corresponds to the propoxy- Ti-O-C vibration and a similar behavior is found for zirconium propoxide with Zr-O-C vibrations located at 1142 cm^{-1}[18].

The acetylacetone vibrational spectrum shows the main features attributed to the most probable staggered conformation of this compound. This conformation has some typical weak IR active vibrations in the low frequency range, as it is shown. Modes at 509, 554 and 640 cm^{-1} correspond to the in plane ring deformation (Δ ring), out of plane ring deformation (Γ ring) and Δ ring + ϱ(CH$_3$) modes, respectively [23].

As described above, when reaction took place, some crystals precipitated short after solution B reached room temperature; IR spectra of these single crystals are also shown in Figure 3. Unlike solution A, a reaction is now verified by the shifting and/or reinforcement of bands associated to reactants vibrations. A discussion of several mechanisms for this kind of reaction has been reviewed by several authors taking into account, primarily, the mixing conditions, the reactivity of metal alkoxides and the Acac/Alkoxides molar ratio [17],[18],[24].

2.4.1. Single crystal XRD characterization

At this stage, a suitable colorless parallelepiped 0.356 mm × 0.162 mm × 0.066 mm was chosen from a representative sample of crystals of the same habit. After the determination of a suitable cell, it was refined by nonlinear least squares and Brava is lattice procedures. The unit cell was then verified by examination of the (h k l) overlays on several frames of data, including zone photographs, and no super cell or erroneous reflections were observed. A search performed on the Cambridge Structural Database and updates using the program Conquest afforded 77 coincidences within 1% of the longest length of a monoclinic C-centered cell whose lattice parameters and reported structure are shown in Figure 4(a). Two entries, those with ACACZR and ZZZADD CCD reference code, in addition, revealed chemical coincidences (both in composition and stoichiometry) for the compound tetrakis(acetylacetonate-O,O')-zirconium(IV) or Zr(Acac)$_4$.

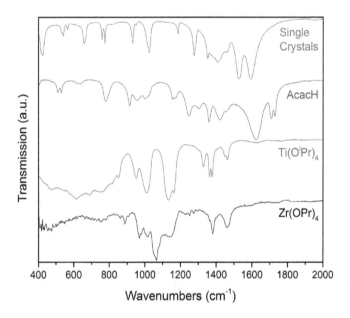

Figure 3. Infrared spectra of the reactants involved in solution B formation. The IR spectra for the resultant precipitated single crystals are also shown.

According to these data, the Zr(Acac)$_4$ structure was simulated using the Accelrys Materials Studio 3D visualizer environment [25]as shown in Figure 4(b). In this case, the acetylacetone reaction with the Zr propoxide results in the formation of an oxo cluster where the metallic atom changes its coordination number from 4 to 8 which is the highest possible value for zirconium. Metallic cations are now bonded to the acetylacetonate chelating ligand giving rise to a less hydrolysable organic complex.

In the case of the titanium propoxide reaction, and taking into account our experimental conditions, it is not a bad assumption to consider the formation of a fully chelated organometallic complex, just as it was described for zirconium. Given the maximum coordination number of 6 for titanium, a Ti(Acac)$_2$(OPr)$_2$ compound will be the most likely to expect as is also confirmed by earlier reports that consider the multiple chelation routes for titanium propoxide [18],[20]. Figure 4(c) shows a very simple Materials Studio 3D modeling for the most probable configuration of $Ti(Acac)_2(OPr)_2$ as determined by the Forcite package [25] on a single step relaxation and energy minimization routine.

Accordingly, bands located at 1280, 1370-1440 and 1530-1595 cm^{-1} in the crystals spectrum, shown in Figure 3, can be assigned to the $v(C-CH_3:Acac)$, $\delta(CH_3:Acac)$ and $v(C-C) + v(C-O:Acac)$ vibrational modes, respectively [26].

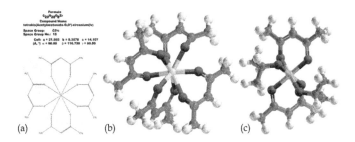

(a)   (b)   (c)

Figure 4. Fully chelated metal-acetylacetonate complexes. (a) $Zr(Acac)_4$ as reported by single crystal XRD charateriza-tion, (b) $Zr(Acac)_4$ structure simulated by Materials Studio (MS) according to its crystallographic data and (c) $Ti(Acac)_2(OPr)_2$ simulated by MS according to its most probable configuration.

2.4.2. Chemical reaction

As seen before, solution B can be thought as the resulting chelated metal complexes mixed with residual isopropanol. In this case, an appropriate chemical equation for the reaction could be:

$$0.53Zr\left(O^iPr\right)_{4(Aq)} + 0.47Ti\left(O^iPr\right)_{4(Aq)} + 2AacH_{(Aq)} \xrightarrow[t\sim4hrs.]{T\sim90^\circ C} 2HO^iPr_{(Aq)} + yZr\left(Acac\right)_{4(s)} +$$
$$+\left(1-2y\right)Ti\left(Acac\right)_2\left(O^iPr\right)_{2(Aq)} + \left(0.53-y\right)Zr\left(O^iPr\right)_{4(Aq)} + \left(2y-0.53\right)Ti\left(O^iPr\right)_{4(Aq)} \quad (2)$$

where we have assumed that all the acetylacetone reacts with the alkoxides thus forming the chelated organometallic complexes. As we see, there is still a fraction of nonchelated reac-tants due to the insufficient amount of acetylacetone needed for that purpose. Let us find now the exact amount of acetylacetone required for obtaining fully chelated organometallic complexes without any further byproducts. We could rewrite equation [2] as:

$$0.53Zr\left(O^iPr\right)_{4(Aq)} + 0.47Ti\left(O^iPr\right)_{4(Aq)} + \left(2+\Delta\right)AcacH_{(Aq)} \xrightarrow[t\sim4hrs.]{T\sim90^\circ C} \left(2+\Delta\right)HO^iPr_{(Aq)} +$$
$$+0.53Zr\left(Acac\right)_{4(s)} + 0.47Ti\left(Acac\right)_2\left(O^iPr\right)_{2(Aq)} \quad (3)$$

Now, it is obvious that $\Delta = 1.06$ implying an alkoxides: acacH molar ratio of 1:3.06 for full chelation of metallic centers. Equation [3] can now be written for any given Zr:Ti ratio (1-x)/x:

$$(1-x)Zr\left(O^i Pr\right)_{4(Aq)} + xTi\left(O^i Pr\right)_{4(Aq)} + 2(2-x)AcacH_{(Aq)} \xrightarrow[t \sim 4hrs.]{T \sim 90^\circ C} 2(2-x)HO^i Pr_{(Aq)} +$$
$$+(1-x)Zr\left(Acac\right)_{4(S)} + xTi\left(Acac\right)_2\left(O^i Pr\right)_{2(Aq)}$$

(4)

This equation represents an extension for the initially proposed sol-gel based route in order to obtain fully chelated and no hydrolysable precursor solutions for PZT based ferroelectric materials. It allows us to synthesize PZT at any Zr:Ti ratio while maximizing solution stability and offering a noticeable repetitiveness for both small and large scale manufacturing and processing.

On the following section we will analyze some features of different sol-gel based PZT (1-x)/x precursors when synthesized using the universal 1:2(2-x) alkoxides: acacH molar ratio already discussed above.

3. Some properties of the PZT precursors

In the final step of the synthesis, when mixing Solutions A and B with 2-methoxyethanol, the dissolved lead acetate complex can react with 2-MOE forming a very stable acetate–methoxyethoxy lead complex [27] that, along with the chelated metal complexes already formed, may lead to turn this final solution into a hydrophobic sol, poorly hydrolysable and, therefore, very stable.

Stability issues tend to be crucial when using sol-gel based precursors in research, small scale applications and industry. Therefore, it is very important to keep control over some parameters that affects the solution stability and, most of all, the average particles size. Particularly, particles size can be an indicator of some undesirable processes that could be taking place in the sol: aggregation, flocculation and sedimentation; aggregation, even though is a reversible process, is a good indicator of instability since the other two processes, which are not reversible, generally follows after some time.

In this respect, the aging time dependences of some physical parameters directly related to the stability of colloidal dispersions must be explored and discussed. Two of the most important parameters that should be taken into account are pH and the average particles size. The first one is determinant for fixing the thickness of the ionic layers surrounding any given charged colloidal particle (Stern layers) while affecting, at the same time, the Zeta potential, the electrophoretic mobility and the aggregation mechanisms as well as the apparent hydrodynamic particle size. As a consequence, the average particles size is the final result of the conjugate action of all the physicochemical variables hardly mentioned before. It is also the definitive experimental variable on which any post processing technique should be based on as it explicitly defines the size range of the so called "building blocks" for bottom-up studies or applications.

Currently, there are no known extensive reports in the literature directly concerned to the study of the time dependence of the average particles size or pH in a PZT precursor sol. It must be noted that, in the case of magnetic nanoparticles, some studies have been carried out [28]-[30]and all of them stress the strong correlation between the size of the colloidal particle and the final properties of the conceived structure, whether it will be nano- or not.

3.1. pH

For this study, we synthesized PZT (1-x)/x precursor solutions by using the same route described in 2.1 with the only difference being the propoxides: acacH molar ratio. In this case, and for the rest of our text, that ratio will be 1:2(2-x) for full chelation of the metal alkoxides.

As it was stated earlier, we will avoid again showing and discussing irrelevant data of samples that showed no significant discrepancies between each other and, because of this, this section will analyze the pH vs. Concentration vs. aging time behaviors of several PZT 53/47 precursor solutions with concentrations ranging from 0.05 M to 0.37 M in a time interval of up to 4 months of stocking.

As can be seen in Figure 5, the acidity/basicity of the solutions clearly shows a tendency with both sols concentration and aging time. Figure 5(a) shows the measured pH vs concentration for the studied sols at several aging steps and, as expected, the more concentrated sample implies the more acidic medium which is consistent with the chemical reaction proposed for sols synthesis.

Figure 5(b), on the other hand, shows the measured pH vs. time behavior. As a general tendency, sols pH dependency with aging time can be divided in three stages which are highlighted in the graph: (i) pH increases notably in the first days after which (ii) it decreases to values somehow close to the initial ones. In this moment, (iii) pH starts to rise again but with a slower time gradient than on stage (i). In our opinion, for the understanding of the pH vs. aging time behavior, we have to take into account the coexistence of different competitive processes right after sols were prepared. In this way, a qualitative description could be done as follows [31]: (i) Remnant unreacted acetic acid evaporates during the first days implying a decrement on the acidity of sols. At this point, (ii) particles are less positively charged and are able to aggregate via condensation reactions followed by the formation of small portions of alcohol (1-propanol) and thus implying a more acidic environment as seen in Figure 5. The number of polyanions per aggregate chain must be limited, however, by the high chemical stability of the chelated metal complexes and that is why the aggregation process does not imply polymerization and/or gelation as it has been reported for more hydrophilic sols. Shortly after condensation rate vanishes, (iii) it is possible for the residual alcohols to evaporate as solutions age thus allowing pH to rise slowly, as seen in Figure 5 for the more aged solutions.

At this scenario, however, hydrolysis is not expected due to (a) the complete chelation of the metal complexes, (b) to the short lengths of the already formed polymeric chains and, therefore, (c) to the small amount of alcohol that could be formed afterwards. After 4 months of aging, no noticeable changes in pH values were detected. Moreover, the stability of these solutions could be eye inspected by verifying neither the absence of sedimentation nor precipitation of single particles or aggregates after almost 1 year of stocking.

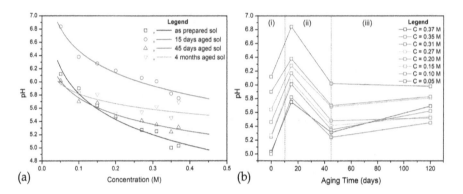

Figure 5. a) pH vs. concentration dependencies for the PZT 53/47 samples understudy measured at different time intervals. Semi-log fittings to guide the eye are also drawn. (b) Aging effects on pH for the same samples. Three distinctive stages were detected and are highlighted in the graph.

3.2. Mean particle size

3.2.1. Experimental technique

Particles size measurements were carried out by means of the dynamic light scattering (DLS) technique in a Zetasizer Nano ZS90 manufactured by Malvern Instruments Ltd. equipped with a HeNe laser source. Approximately 1.5 ml of as synthesized sols were stored in polystyrene cells (DTS0012 cells) provided by the same manufacturer and then size distributions curves were recorded from the very first day until 125 days after synthesis. For this purpose, several solvent parameters were needed for further processing of dispersed light intensity, namely viscosity ($\eta_{2\text{-MOE}} \sim 1.5410$ cP), dielectric constant ($\varepsilon_{2\text{-MOE}} \sim 16.9$) and refraction index ($n_{2\text{-MOE}} \sim 1.33$) [32]. After each measurement, a nonlinear least squares fitting (NLLSF) was made to the experimental data according to a log-normal distribution function:

$$f_{\ln}(x, \beta_i) = \beta_0 + \frac{\beta_1}{\sqrt{\pi}\beta_2} \frac{1}{x} \exp\left[-\frac{1}{2\beta_2^2} \ln\left(\frac{x}{\beta_3}\right)^2 \right] \tag{5}$$

Where β_i are fitting parameters. It must be stressed that this fitting was carried out only for size intervals where unimodal distribution curves were measured.

3.2.2. Concentration dependence

Figure 6 shows the results obtained for particles size distribution for two concentrations under study (a PZT 53/47 precursor) after different time lapses that are also shown in the figure.

According to what is shown, an appropriate discussion relaying on the acidity/particles size dependence cannot be easily established: oscillations in the pH values, see Figure 5, during aging are not followed by oscillations in the mean particles size values. Even though this feature is not fully understood, it may be due to the short lengths of the oligomeric chains present in the solution and to the weak sensitivity of the completely chelated metal complexes to the ionic strength of the solvent medium. Another possible explanation could be given in terms of the irreversibility associated to the formation of these chains when the solvent medium is not acidic enough to break the corresponding bonds and/or coordinations.

As can be seen, as prepared sols featured multimodal distributions with some small percent of particles that could even have sizes higher than 1μm. As solutions are aged, these distributions tend to be unimodal featuring a mean particles size well below 10 nm. We consider that this is a remarkable result of this work as it shows that there is no rule of thumb closely related to the need of using fresh, as prepared solutions for synthesizing several kinds of nanodimensional systems by means of electrophoretic deposition(EPD) or simple template immersion, [33]-[36]. Moreover, in this case we could have chosen solutions aged for 20 days or more for bulk, thin films or nanostructures synthesis due to the better homogeneity regarding particles size distribution in both cases under study.

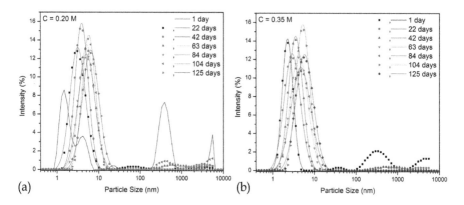

Figure 6. Particles size distribution functions for the analysed PZT 53/47 precursor solutions and for several aging times in the whole measurement range. Log-normal fittings for the more populated size intervals are also depicted.

This behavior could be explained, in principle, on the basis of the chemical reaction which was described for obtaining what we called solution B, see Section 2.4. As we stated, by the end of this step we observed the crystallization of the compound $Zr(Acac)_4$. Such crystallization does not occur after mixing solutions A and B in the presence of 2-MOE which can be attributed (I) to the higher acidity of the final A+B+2-MOE solution, (II) to the solubility of

Zr(Acac)$_4$ in 2-MOE, (III) to a dilution process while mixing A+ B, or (IV) to a combined effect of all of the above. Must be noted that the less acidic solution (C = 0.20M) possesses, at the same time, the less unimodal distribution function which is an expected result in concordance to the relationship between ionic strength and particles size.

Thus, it seems plausible to assume solution B as a dispersion containing the Ti(Acac)$_2$(OPr)$_2$ compound at molecular level and submicrometric aggregates of Zr(Acac)$_4$ that do not precipitate; these are redissolved afterwards when mixed with the acetate rich solution A and 2-MOE. After stirring the resultant sol for one day, the size of some percentage of the Zr(Acac)$_4$ aggregates still ranges between 100 and 1000 nm; after 20 days, however, size is well below 10 nm. From these results, and somehow confirming our previous discussion, we can also see that the less concentrated solution features a considerable amount of particles with sizes well above 50nm even after 100 days of stocking. For this reason, we will focus our attention from now on in precursor solutions with 0.35 M aged during the first 30-35 days, a time period that, according to our results, seems to be critical in the colloidal stability of as synthesized sols.

3.2.3. Aging time dependence

As it was said before, all samples under study featured a noticeable stabilization when aged for about 1 month. Figure 7 shows this aging behavior for two PZT (1-x)/x precursor solutions concentrated at 0.35 M. In all cases, it could be observed a similar time evolution as the one described earlier for Figure 6.

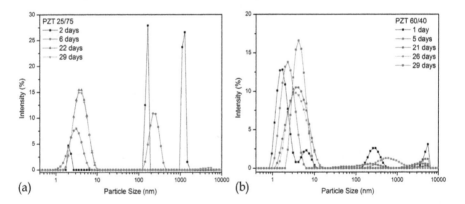

Figure 7. Aging behavior, as depicted by the time evolution of particles size distribution curves, for two PZT (1-x)/x precursor solutions concentrated at 0.35 M: (a) PZT 25/75 and (b) PZT 60/40.

On the other hand, Figure 8 shows the whole dataset of the measured mean particles size for the synthesized solutions. Fitting curves shown here were determined by considering a time dependence given by $<d> = C_1 t^{1/(1-\lambda)} + C_2$ that, according to the classical Smoluchowski theory in the van Dongen–Ernst approximation, describes appropriately a nongelling system of dispersed clusters. Moreover, in the same approximation, the small values of $\lambda(\lambda \ll 1)$ are usually associated with a Brownian diffusion limited aggregation process vastly dominated by collisions between larger clusters with smaller ones [31].

Another view of the aging process could be illustrated by means of the dimensionless normalized distribution curves, not shown here. In that representation, the broadening of the distribution curves with aging is an indicator of very likely aggregation processes and somehow will help to complete the kinetic analysis that we have been through in this section.

In a way of summarizing our results, it could be said that, just after synthesis, several populations of particles were detected until homogenized a few days later; then particles tend to grow very slowly with time according to an almost linear law (corresponding to that describing a nongelling system) and aggregation is expected. Moreover, these cases fit very well in the Brownian diffusion limited aggregation type where smaller clusters stick to bigger ones when they collide. The slow but evident increase in the mean cluster size, as well as the noticeable broadening in the distribution peaks, must warn us about an undesirable and irreversible precipitation process for, hypothetically, $t \rightarrow \infty$.

The high stability shown by our samples even after 4 months of stocking and the small values for mean particles size (well below 10 nm) are good indicators that the complete chelation of the organometallic compounds plays a key role in keeping short, and poorly reactive, oligomeric chains. An aggregation process dominated exclusively by Brownian motion is highly desirable when solution stability needs to be maximized.

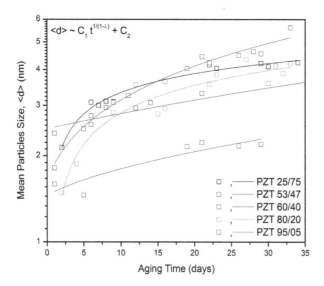

Figure 8. Mean particles size vs. aging time dependence for the systems under study. Fitting curves obeying a $t^{1/(1-\lambda)}$ scaling law are depicted.

Our samples were also characterized by the High Resolution Transmission Electron Micro-scopy (HR-TEM) technique. A few drops of the synthesized sols were deposited on copper grids, evaporated afterwards and put under a JEOL JEM2200 microscope with Omega filter and a spherical aberration corrector; some of the obtained images are shown in Figure 9 for a PZT 95/05 precursor solution. As expected, particles are likely to coalesce as the solvent evaporated prior to characterization, see Figure 9(a) and (b), and some of those nanoparti-cles are shown in higher magnification images in Figure 9(c) and (d). Most of the features regarding particle size that were discussed earlier were corroborated by means of this tech-nique for all the precursor solutions.

At this point, and considering what has been discussed until now, it must be highlighted that a rigorous control of the chelation rate, pH and aging time could give us a chance to "tune" the average nanoparticles size and/or the colloidal stability as desired. By looking back at Figure 5 through Figure 8 one may feel free to choose to explore several "working points" according to our research or technological needs for every (1-x)/x Zr:Ti ratio in the PZT system.

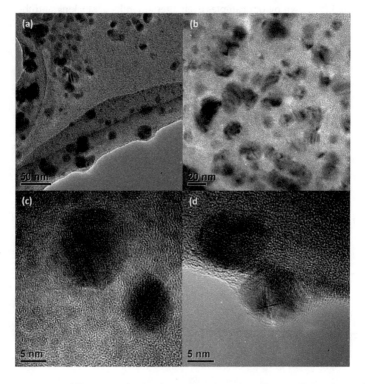

Figure 9. HR-TEMimages at different magnifications for a PZT 95/05 precursor solution, with C = 0.35 M and aged for 2 months.

This, in fact, is a powerful tool provided by the physicochemical phenomena and mechanisms previously discussed and, generally speaking, could be extended to other material systems of technological interest at a moderate cost.

4. Crystallization route

The final A + B + 2-MOE solutions were dried at 100°C for several days and, after that, they were thermally treated in order to analyze the phase evolution from the amorphous PZT sol-gel network to the expected final Perovskite structure. Due to fundamental similarities among the different samples we will be discussing only the case where Zr:Ti ratio is 53/47, concentrated at 0.35 M and aged for 2 months.

4.1. Experimental techniques

Crystallization was monitored by means of FT-IR and Raman vibrational spectroscopies, see Sections 2.2.1 and 2.2.2, and by X-Ray Diffraction (XRD) and Scanning Electron Microscopy (SEM) techniques after treating the samples for 12 hours at certain temperatures that were previously chosen. For the sake of clarity, we decided to divide the crystallization study in two temperature intervals: (i) $100 \leq T \leq 510°C$ and (ii) $550 \leq T \leq 900°C$. At this point, it is important to say that, for powders heated at 850°C and 900°C, the treatment was carried out for just 2 hours due to the high volatility of lead for 850°C and beyond.

Scanning Electron Microscopy (SEM) was carried out on fine ground powders in a Leica Cambridge Stereoscan 440 microscope and the X-Ray powder diffraction patterns were recorded over a 20-60° 2θ range on a Bruker D8 Advance diffractometer with filtered CuKα radiation. In this technique, the identified phases were indexed by comparing the resulting diffraction patterns with those of similar compositions reported in the International Union of Crystallography (IUCr) JCPDS-ICDD database.

4.2. $100 \leq T \leq 510$ °C

In this temperature range the material goes through very noticeable phase transformations; the initial amorphous powder starts to show some crystallinity for about 500°C, right after the combustion of the remaining organic species. Figure 10 shows these first crystallization stages as recorded by the Raman (a) and FT-IR (b) vibrational spectroscopies. The vibrational modes associated to the remnant organic compounds coexisting after heat treating at 100 °C are also highlighted in this graph, Figure 10(b), and it must be noticed that they correspond, basically, to the lead acetate and to the completely chelated organometallic complexes. More explicitly, numbered modes in Figure 2(b) have been assigned to: (1)$Pb(OAc)_2$:$\varrho(COO)$, (2)$Pb(OAc)_2$:$\delta_S(COO)$, (3)$Pb(OAc)_2$:$\nu_S(C\text{-}C)$, (4)$Ti(OPr)_2(Acac)_2$:Ti-Acac, (5)$Zr(Acac)_4$:Zr-Acac,(6)$Pb(OAc)_2$:$\delta_S(CH_3)$, (7)$Ti(OPr)_2(Acac)_2/Zr(Acac)_4$:$\delta(CH_3$:Acac) + $Pb(OAc)_2$:$\nu_S(CO)$and (8)$Ti(OPr)_2(Acac)_2/Zr(Acac)_4$:[$\nu(C\text{-}C$ + $\nu(C\text{-}O$:Acac)] + $Pb(OAc)_2$:$\nu_A(C=O)$[26],[37].

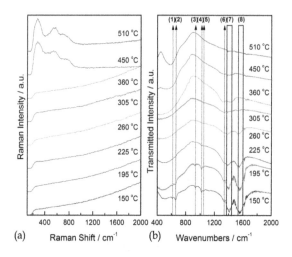

(a) Raman Shift / cm⁻¹   (b) Wavenumbers / cm⁻¹

Figure 10. Phase transformations for 100 °C≤ T≤ 510 °C as registered by Raman (a) and FT-IR (b) vibrational spectroscopies; in this case, remnant vibrations due to the presence of acetate and chelated metal complexes are highlighted.

Until 225°C, IR spectra showed no variation; the intensity ratios between the detected modes
are almost the same and, on the other hand, Raman spectra featured no signal given the
strong presence of π-bonded organic species. Afterwards, IR spectra revealed the almost
complete decomposition of acetates and a decrement of Metal:Acac complexes while any noticeable Raman signal is still absent. Just after T = 450°C the active Raman modes normally
assigned to a PZT 53/47 started to show up. After treating at this temperature, the FT-IR
spectra reveals the formation and definition of the A1(3TO) vibrational mode (~600 cm⁻¹)
which is also characteristic for this compound [38]-[41].

Another view of the whole process in this temperature range could be given, as seen in Figure 11, by means of XRD characterization. XRD patterns clearly illustrate the phase transformation exhibited by an amorphous material turning into crystalline; as for the Raman
spectra, crystallinity starts to get noticed at T = 450°C. However, these patterns revealed the
existence of well defined peaks near 28° that disappear almost entirely at 510°C.

These maxima are usually associated to an intermediate metastable phase belonging to the
Fluorite crystal system (F) which is mainly characterized by the spatial disarrangement of
the oxygen atoms, vacancies and metal cations while coexisting with some carbonates
and/or oxides that still remain in the material. Afterwards, when carbonates and oxides react, a new metastable phase is formed but this new arrangement tends to be more ordered
than the previous fluorite. This intermediate phase is considered as belonging to the Pyrochlore crystal system (P or Pyr) and consisting of a 2 x 2 x 2 ordered fluorite cell with oxygen
and metallic vacancies or, from another point of view, consisting of a 2 x 2 x 2 non stoichiometric Perovskite (Per) cell. This rigorous differentiation between Fluorite and Pyrochlore is
not always considered in literature and it can be subtle sometimes, especially when working

with a well known compound as PZT is. Anyway, in Figure 11 we have denoted by F (Fluorite) and P (Pyrochlore) the diffraction maxima associated with each of these phases and according to the indexed cubic structures reported in reference [42] (a = 5.25 Å) and in reference [43] (a = 10.48Å), respectively.

Up to this point, we have seen the almost complete elimination of organic ligands with heating as well as the amorphous/crystalline phase transformation going through two intermediate metastable phases. In the next subsection we will carry on a similar study for higher temperatures while exploring the formation of pure Perovskite phase with morphological quality.

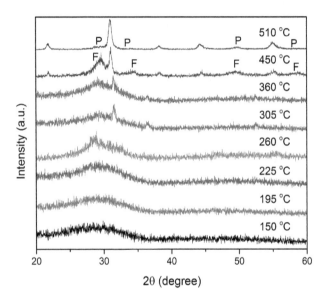

Figure 11. Thermal evolution of PZT 53/47 precursor powders when 100 °C≤ T≤ 510 °C as recorded by XRD patterns.

4.3. 550≤ T ≤ 900 °C

Figure 12 shows Raman (b) and FT-IR (b) spectra for heat treated powders in the 550 – 900 °C range. Unlike the previously studied temperature range, vibrational spectra did not show drastic or very noticeable changes. The FT-IR spectra features the only IR active band for aPZT-R3m in the 400-1500 cm^{-1} range (A1(3TO)); this band, associated to the extensional vibrations of the Perovskite BO$_6$ octahedra, shifts slightly to higher frequencies while gets narrower as temperature increases. This shifting suggests a more compact octahedral structure while a narrower band could be the evidence of a better 'environment' around the octahedron or, in other words, a better local stoichiometry [44] that seems to be minimum for T = 800°C, an indicative of an optimum crystallization. Raman spectra featured vibrational modes that tend to define better as temperature increases.

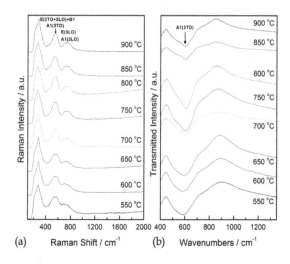

Figure 12. Phase transformations for 550 °C≤ T≤ 900 °C as registered by Raman (a) and FT-IR (b) vibrational spectroscopies.

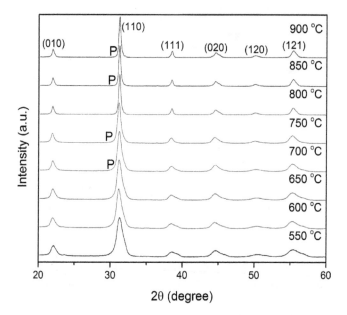

Figure 13. Thermal evolution of the powders under study when 550 °C ≤ T ≤ 900 °C as evidenced by XRD patterns in the 20 - 60 deg. 2θ interval. Regions where the most intense Pyr peak showed up are denoted by P.

XRD patterns depicted in Figure 13 allow us to notice the temperature evolution of the distinctive Perovskite diffraction peaks. Peaks indexing has been done according to a PZT 53/47, $F_{R(HT)}$; sym: R3m. Besides, regions where the most intense Pyr peak showed up (~28.5°) are denoted by P; the Pyr phase is present in a small percentage (%Pyr ~ 1 %) except for powders treated at T = 800°C which must be chosen as the appropriate crystallization temperature for this material. The final Perovskite phase homogenizes and tends to be predominant as temperature increases until lead losses become noticeable.

The rest of the PZT compositions that were chosen for this study, PZT (1-x)/x with C = 0.35 M and 2 months aged, were also treated at 800 °C in order to obtain the appropiate Perovskite phase; Figure 14 shows the corresponding XRD patterns for each one of them. As seen in the graph, all samples attained a perfect crystallization in the corresponding phase and the proper pattern indexing as well as the phase indentification are also shown.

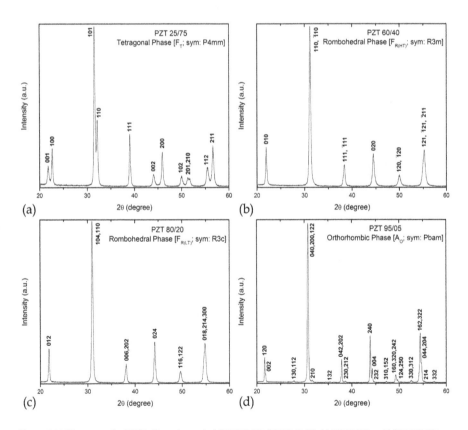

Figure 14. XRD patterns for PZT (1-x)/x under study: (a) PZT 25/75, (b) PZT 60/40, (c) PZT 80/20 and (d) PZT 95/05.

82

Handbook of Ferroelectrics

On the other hand, Figure 15 shows SEM micrographs of crystallized PZT (1-x)/x powdered samples. As illustrated, a granular structure of fair morphological uniformity is found in all cases along with a remarkable submicrometric average grain size. Another interesting feature is the low porosity found in all compositions, somehow resembling the granular structure of sintered bulk samples, which certainly favors the formation of a high density ferroelectric material.

As it was said at the beginning of this chapter, this kind of sub-micron granular structures, or nanoceramics, allows the technological exploitation of dielectric and ferro/piezoelectric size related features. It is not shown here, but that very same samples have been resynthesized a couple of times with starting reactants being bought to different companies and final results showed almost the same granular and morphological quality. Repetitiveness is also a bonus when working with chemical routes of synthesis that tend to be less straightforward, even though cheaper, than physical ones.

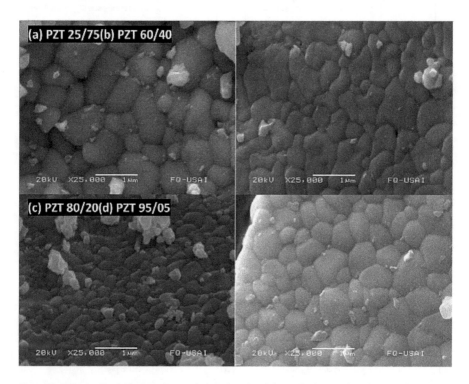

Figure 15. SEM micrographs showing the morphological quality of the PZT (1-x)/x powders when crystallized at 800 °C: (a) PZT 25/75, (b) PZT 60/40, (c) PZT 80/20 and (d) PZT 95/05.

5. Chapter remarks

Through this chapter, we tried to expose some relevant results directly concerned with a fea-
sible 'universal' controlled synthesis of nano/submicrometric grain-sized PZT (1-x)/x piezoelectric structures that, technically speaking, potentially enhance the performance of current
commercially bulk based devices by exploiting the size effects related phenomena that arise,
as is commonly accepted, for grain sizes below 1 μm. Under the light of this study, the tenta-
tive bottom-up "design" of any desired nano/submicrometric PZT (1-x)/x structure seems to
be a plausible and successful task that, methodologically at least, could be expanded to more
complex materials systems.

Further reading regarding the work that has been shown here can be found in references
[45]-[48]. On the other hand, references [49]-[60] will also provide the reader with very recent research papers in this field that undoubtedly extend the applicability and versatility of
what has been discussed.

Author details

A. Suárez-Gómez[1], J.M. Saniger-Blesa[2] and F. Calderón-Piñar[3]

1 UdG-CUVALLES, Carr. Guadalajara-Ameca, Ameca, Jalisco, México

2 CCADET-UNAM, Cd. Universitaria, Coyoacán, México D.F., México

3 Fac. de Física/IMRE, San Lázaro y L, Univ. de la Habana, Habana , Cuba

References

[1] McCauley D, Newnham RE, Randall CA. Intrinsic size effects in a barium titanate
glass-ceramic. Journal of the American Ceramic Society1998;81(4)979-987.

[2] Frey MH, Xu Z, Han P, Payne DA. The role of interfaces on an apparent grain size
effect on the dielectric properties for ferroelectric barium titanate ceramics. Ferroelec-
trics1998;206(1)337-353.

[3] Zhang L, Zhong LW, Wang CL, Zhang PL, Wang YG. Finite-size effects in ferroelectric solid solution $Ba_xSr_{1-x}TiO_3$. Journal of Physics D: Applied Physics 1999;32 546-551.

[4] Park Y, Knowles KM, Cho K. Particle-size effect on the ferroelectric phase transition
in $PbSc_{1/2}Ta_{1/2}O_3$ ceramics. Journal of Applied Physics 1998;83(11)5702-5708.

[5] Desu S.B., Ramesh R., Tuttle B.A., Jones R.E., Yoo I.K., editors. Ferroelectric Thin
Films V (MRS Symposium Proceedings Series Vol. 433), San Francisco, CA, 1996.

[6]   Wang ZL. The new field of nanopiezotronics. Materials Today 2007;10(5) 20-28.

[7]   Budd K.D., Dey S.K., Payne DA. Sol-Gel Processing of $PbTiO_3$-$PbZrO_3$ and PLZT Thin Films. Proceedings of the British Ceramic Society 1985;36107–121.

[8]   Schneller T, Waser R. Chemical Solution Deposition of Ferroelectric Thin Films: State of the Art and Recent Trends. Ferroelectrics 2002;267(1) 293–301.

[9]   Majumder SB, Bhaskar S, Katiyar RS. Critical issues in sol-gel derived ferroelectric thin films: A review. Integrated Ferroelectrics 2002:42(1)245–292.

[10] Besra L, Liu M. A review on fundamentals and applications of electrophoretic deposition (EPD). Progress in Materials Science 2007;52(1) 1-61.

[11] Shantha Shankar K, Raychaudhuri AK. Fabrication of nanowires of multicomponent oxides: Review of recent advances. Materials Science and Engineering: C 2005;25(5–8) 738-751.

[12] Ogihara T, Kaneko H, Mizutani N, Kato M. Preparation of monodispersed lead zirconate-titanate fine particles. Journal of Materials Science Letters 1988;7(8) 867-869.

[13] Lakeman CDE, Payne DA. Processing Effects in the Sol–Gel Preparation of PZT Dried Gels, Powders, and Ferroelectric Thin Layers. Journal of the American Ceramic Society 1992;75(11) 3091-3096.

[14] Wang FP, Yu YJ, Jiang ZH, Zhao LC. Synthesis of $Pb_{1-x}Eu_x(Zr_{0.52}Ti_{0.48})O_3$ nanopowders by a modified sol–gel process using zirconium oxynitrate source. Materials Chemistry and Physics 2003;77(1)10-13.

[15] Brunckova H, Medvecky L, Briancin J, Saksi K. Influence of hydrolysis conditions of the acetate sol–gel process on the stoichiometry of PZT powders. Ceramics International 2004;30(3) 453-460.

[16] Praveenkumar B, Sreenivasalu G, Kumar HH, Kharat DK, Balasubramanian M,Murty BS. Size effect studies on nanocrystalline $Pb(Zr_{0.53}Ti_{0.47})O_3$ synthesized by mechanical activation route. Materials Chemistry and Physics 2009;117(2–3) 338-342.

[17] Livage J, Henry M, Sanchez C. Sol-gel chemistry of transition metal oxides. Progress in Solid State Chemistry 1988;18(4)259–342.

[18] Sedlar M, Sayer M. Reactivity of titanium isopropoxide, zirconium propoxide and niobium ethoxide in the system of 2-Methoxyethanol, 2,4-Pentanedione and water. Journal of Sol–Gel Science and Technology 1995;5(1) 27–40.

[19] Nakamoto K. Infrared and Raman Spectra of Inorganic and CoordinationCompounds. John Wiley & Sons; 1997.

[20] Doeuff S, Henry M, Sanchez C, Livage J. Hydrolisis of titanium alkoxides: Modification of the molecular precursor by acetic acid. Journal of Non-Crystalline Solids 1987;89(1-2) 206–216.

[21] Lewandowski H, Koglin E, Meier RJ. Computational study of the infrared spectrum of acetic acid, its cyclic dimer, and its methyl ester. Vibrational Spectroscopy 2005;39(1) 15–22.

[22] Yang MM, Crerar DA, Irish DE. A Raman spectroscopic study of lead and zinc acetate complexes in hydrothermal solutions. Geochimica et Cosmochimica Acta 1989;53(2) 319–326.

[23] Tayyari SF, Milani-nejad F. Vibrational assignment of acetylacetone. Spectrochimica Acta A: Molecular and Biomolecular Spectroscopy2000;56(14) 2679–2691.

[24] Schubert U, Husing N, Lorenz A. Hybrid Inorganic-Organic Materials by Sol-Gel Processing of Organofunctional Metal Alkoxides. Chemistry of Materials 7(11) 2010–2027.

[25] Accelrys Inc., San Diego, CA USA. http://accelrys.com/products/materials-studio/

[26] Schwartz RW, Voigt JA, Boyle TJ, Christenson TA, Buchheit CD,"Control of Thin Film Processing Behavior Through Precursor Structural Modifications. Ceramic Engineering and Science Proceedings 1995;16(5)1045–1056.

[27] Kolb U, Gutwerk D, Beudert R, Bertagnolli H. An IR- and EXAFS-study of the precursor system lead(II) acetate trihydrate, dissolved in methanol and 2-methoxyethanol. Journal of Non-Crystalline Solids 1997;217(2-3) 162-166.

[28] Zeisgerber M, Dutz S, Lehnert J, Müller R. Measurement of the distribution parameters of size and magnetic properties of magnetic nanoparticles for medical applications. Journal of Physics: Conference Series 2009;149(1) 012115.

[29] Ortega D, Garitaondia JS, Ramírez-del-Solar M, Barrera-Solano C, Domínguez M. Implications of nanoparticle concentration and size distribution in the superparamagnetic behaviour of aging-improved maghemite xerogels. The European Physics Journal D: Atomic, Molecular, Optical and Plasma Physics 2009;52(1-3) 19-22.

[30] Maity D, Choo SG, Yi J, Ding J, Xue JM. Synthesis of magnetite nanoparticles via a solvent-free thermal decomposition route. Journal of Magnetism and Magnetic Materials 2009;321(9) 1256-1259.

[31] Brinker CJ, Scherer GW. Sol–gel Science: The Physics and Chemistry of Sol–gel Processing. Academic Press Inc., San Diego; 1990.

[32] Das B, Hazra DK. Conductance of selected Alkali Metal Salts in Aqueous Binary Mixtures of 2-Methoxyethanol at 25 °C. Journal of Solution Chemistry 1998;27(11) 1021-1031.

[33] Valdés-Solís T, Marbán G, Fuertes AB. Preparation of Nanosized Perovskites and Spinels through a Silica Xerogel Template Route. Chemistry of Materials 2005;17(8) 1919-1922.

[34] Steinhart M, Jia Z, Schaper AK, Wehrspohn RB, Gösele U, Wendorff JH. Palladium Nanotubes with Tailored Wall Morphologies. Advanced Materials 2003;15(9) 706-709.

[35] Limmer SJ, Hubler TL, Cao G. Nanorods of Various Oxides and Hierarchically Structured Mesoporous Silica by Sol-Gel Electrophoresis. Journal of Sol–Gel Science and Technology 2003;26(1-3) 577-581.

[36] Urban JJ, Yun WS, Gu Q, Park H. Synthesis of Single-Crystalline Perovskite Nanorods Composed of Barium Titanate and Strontium Titanate. Journal of the American Chemical Society 2002;124(7) 1186-1187.

[37] Hardy A, Van Werde K, Vanhoyland G, Van Bael MK, Mullens J, Van Poucke LC. Study of the decomposition of an aqueous metal–chelate gel precursor for $(Bi,La)_4Ti_3O_{12}$ by means of TGA–FTIR, TGA–MS and HT-DRIFT. Thermochimica Acta 2003;397(1-2)143-153.

[38] Petzelt J, Ostapchuk T. Infrared and Raman spectroscopy of some ferroelectric perovskite films and ceramics. Journal of Optoelectronics and Advanced Materials 2003;5(3) 725-733.

[39] Meng JF, Katiyar RS, Zou GT, Wang XH. Raman Phonon Modes and Ferroelectric Phase Transitions in Nanocrystalline Lead Zirconate Titanate. Physica Status Solidi A: Applications and Materials Science 1997;164(2)851-862.

[40] Burns G, Scott BA. Raman Spectra of Polycrystalline Solids; Application to the $PbTi_{1-x}Zr_xO_3$ System. Physical Review Letters 1970;25(17) 1191-1194.

[41] Camargo ER, Leite ER, Longo E. Synthesis and characterization of lead zirconate titanate powders obtained by the oxidant peroxo method. Journal of Alloys and Compounds 2009;469(1-2)523-528.

[42] Wilkinson AP, Speck JS, Cheetham AK, Natarajan S, Thomas JM. In Situ X-ray Diffraction Study of Crystallization Kinetics in $PbZr_{1-x}Ti_xO_3$ (PZT, x = 0.0, 0.55, 1.0). Chemistry of Materials 1994;6(6)750-754.

[43] Kwok CK, Desu SB. Pyrochlore to perovskite phase transformation in sol-gel derived lead-zirconate-titanate thin films. Applied Physics Letters 1992;60(12)1430-1432.

[44] Lakeman CDE, Xu Z, Payne DA. Rapid Thermal Processing of Sol-Gel Derived PZT 53/47 Thin Layers. Proceedings of the Ninth IEEE International Symposium onApplications of Ferroelectrics, 1994. ISAF '94. 1995404-407.

[45] Suárez-Gómez A, Sato-Berrú R, Toscano RA, Saniger-Blesa JM, Calderón-Piñar F. On the synthesis and crystallization process of nanocrystalline PZT powders obtained by a hybrid sol–gel alkoxides route. Journal of Alloys and Compounds 2008;450(1-2) 380-386.

[46] Suárez-Gómez A, Saniger-Blesa JM, Calderón-Piñar F. The effects of aging and concentration on some interesting Sol-gel parameters: A feasibility study for PZT nano-

particles insertion on in-house prepared PAA matrices via electrophoresis. Journal of Electroceramics 2009;22(1-3) 136-144.

[47] Suárez-Gómez A, Saniger-Blesa JM, Calderón-Piñar F. A study on the stability of a PZT precursor solution based on the time evolutionof mean particles size and pH. Materials Chemistry and Physics 2010;123(1) 304-308.

[48] Suárez-Gómez A, Saniger-Blesa JM, Calderón-Piñar F. A Crystallization Study of Nanocrystalline PZT 53/47 Granular Arrays Using a Sol-Gel Based Precursor. Journal of Materials Science & Technology 2011;27(6) 489-496.

[49] Meng Q, Zhu K, Pang X, Qiu J, Shao B, Ji H. Sol-hydrothermal synthesis and characterization of lead zirconate titanate fine particles. Advanced Powder Technology 2012;In Press DOI:10.1016/j.apt.2012.06.004.

[50] Chiolerio A, Quaglio M, Lamberti A, Celegato F, Balma D, Allia P. Magnetoelastic coupling in multilayered ferroelectric/ferromagnetic thin films: A quantitative evaluation. Applied Surface Science 2012;258(20) 8072–8077.

[51] Bastani Y, Bassiri-Gharb N. Enhanced dielectric and piezoelectric response in PZT superlattice-like films by leveraging spontaneous Zr/Ti gradient formation. Acta Materialia 2012;60(3) 1346–1352.

[52] Stawski TM, Besselink R, Veldhuis SA, Castricum HL, Blank DHA, ten Elshof JE. Time-resolved small angle X-ray scattering study of sol–gel precursor solutions of lead zirconate titanate and zirconia. Journal of Colloid and Interface Science 2012;369(1) 184–192.

[53] Goel P, Vijayan N, Biradar AM. Complex impedance studies of low temperature synthesized fine grain PZT/CeO$_2$ nanocomposites. Ceramics International 2012;38(4) 3047–3055.

[54] Yin Y, Ye H, Zhan W, Hong L, Ma H, Xu J. Preparation and characterization of unimorph actuators based on piezoelectric Pb(Zr$_{0.52}$Ti$_{0.48}$)O$_3$ materials. Sensors and Actuators A: Physical 2011;171(2) 332–339.

[55] Sachdeva A, Tandon RP. Effect of sol composition on dielectric and ferroelectric properties of PZT composite films. Ceramics International 2012;38(2) 1331–1339.

[56] Bochenek D, Skulski R, Wawrzała P, Brzezińska D. Dielectric and ferroelectric properties and electric conductivity of sol–gel derived PBZT ceramics. Journal of Alloys and Compounds 2011;509(17) 5356–5363.

[57] Chen W, Zhu W, Chen XF, Sun LL. Preparation of (Ni$_{0.5}$Zn$_{0.5}$)Fe$_2$O$_4$/Pb(Zr$_{0.53}$Ti$_{0.47}$)O$_3$ thick films and their magnetic and ferroelectric properties. Materials Chemistry and Physics 2011;127(1–2) 70–73.

[58] Zhong C, Wang X, Li L. Fabrication and characterization of high curie temperature BiSc$_{1/2}$Fe$_{1/2}$O$_3$–PbTiO$_3$ piezoelectric films by a sol–gel process. Ceramics International 2012;38(S1) S237–S240.

[59] Fuentes-Fernandez E, Baldenegro-Perez L, Quevedo-Lopez M, Gnade B, Hande A, Shah P, Alshareef HN. Optimization of $Pb(Zr_{0.53}Ti_{0.47})O_3$ films for micropower generation using integrated cantilevers. Solid-State Electronics 2011;63(1) 89–93.

[60] Kim SH, Koo CY, Lee J, Jiang W, Kingon AI. Enhanced dielectric and piezoelectric properties of low-temperature processed $Pb(Zr,Ti)O_3$ thick films prepared by hybrid deposition technique with chemical solution infiltration process. Materials Letters 2011;65(19–20) 3041–3043.

Thin-Film Process Technology for Ferroelectric Application

Koukou Suu

Additional information is available at the end of the chapter

1. Introduction

Recently thin-film ferroelectrics such as Pb(Zr, Ti)O$_3$ (PZT) and (Ba, Sr)TiO$_3$ (BST) have been utilized to form advanced semiconductor and electronic devices including Ferroelectric Random Access memory(FeRAM), actuators composing gyro meters, portable camera modules, and tunable devices for smart phone applications and so on. Processing technology of ferroelectric materials is one of the most important technologies to enable the above- mentioned advanced devices and their productions.

In this paper, we will report our development results of ferroelectric thin film processing technologies including sputtering, MOCVD and plasma etching as well as manufacturing processes for FeRAM, MEMS(actuators, tunable devices) and ultra-high density probe memory.

2. FeRAM technology

Ferroelectric Random Access Memory (FeRAM) is a main candidate of next generation non-volatile random access memory. As its density has been continuously increasing, its applications have spread from RFID card to Nonvolatile latch circuit with low power consumption. FeRAM has a ferroelectric thin films sandwiched by rare metal electrodes and we have established mass production capability on integrated FeRAM solution, consisting of electrode film deposition, PZT thin-film deposition, etching/ashing, anneal and passivation deposition.

2.1. Mass productive sputtering technology for perovskite oxide thin-film

We have been developing mass production technology for perovskite oxide thin-film such as PZT for a long period [1-5], since we consider these materials as the most promising candidate for ferroelectric material used in FeRAMs (Ferroelectric Random Access Memories) as a non-volatile memory device, piezoelectric MEMS (Micro Electro Mechanical System) devices due to its longer period of research, existence of actual production, manufacturing capability within the tolerable temperature range of general Si LSI technology.

The sputtering was selected for mass production technology of perovskite oxide thin-films owing to the following factors: (1) Good compatibility with conventional Si LSI processes. (2) Superb controllability of film quality (e.g. film composition), which enabled relatively easy thin film deposition. (3) Better possibilities of obtaining uniform surfaces in large diameter substrates (e.g. 6-8 inch). (4) Sputtering was plasma processing which was promising for deposition and heat treatment at low temperature. (5) Feasibility of high-speed deposition. (6) Same deposition method as electrodes (Pt, Ir, Ru, etc.), which will facilitate in-situ integration. (7) Present difficulties and lack of future potentialities in other technologies.

With emphasis on both mass production capability and advanced process capacity of perovskite oxide thin-film sputtering, we consider the following factors as important for development:

1.   Throughput

Compared to the other processes (e.g. electrode deposition), the deposition speed of perovskite oxide material sputtering was considerably slower and thereby limited the throughput. Though two ways for improving the throughput, high-speed deposition and thinner film deposition, are considered, the former is more promising for the improvement of throughput, while the latter is apt to cause deterioration of film characteristics.

2.   Control of film composition

Film composition determines other film qualities (crystal structure, electric characteristics). Since volatile elements were included in the perovskite oxide materials, they were sensitive and easily fluctuated according to temperature or plasma status. Film composition control is the fundamental factor in this process.

3.   Uniformity over large diameter substrates

Large diameter substrates up to 8 inch were expected to be used for mass production of perovskite oxide thin-film. Film thickness and uniformity of film quality were the keys to mass production.

4.   Process stability / reliability

Under the circumstances where there were many unknown factors such as new materials, ceramic targets, insulator sputtering, etc., process stability / reliability was the more pertinent factor. In fact, there was a problem in film composition that changed over time.

5.   Prevention of particles

While the characteristics, such as ceramic targets or insulating thin film, increased the mechanical factors (e.g. adherence, thermal expansion) for particle occurrence, they were also producing electric factors (e.g. dielectric breakdown due to charge-up) and made the measures difficult to obtain.

2.2. Optimization of sputtering processes for perovskite oxide thin-film

We adapted the RF magnetron sputtering method for perovskite oxide thin-film sputtering. As for the PZT thin-films, sputtering methods included high-temperature deposition, where film deposition was made at substrate temperatures above 500°C [6, 7], and low temperature deposition, where the films were deposited at room temperature and then crystallized by post-annealing process. The improvement in high-speed deposition, film composition control, and stability of sputtering processes is described in the following.

If the ceramic target with inferior thermal conductivity is used, application of high power leads to the destruction of the target. By adopting the backing plate with high cooling efficiency and high-density target, higher deposition rate by an increase in sputtering power is achieved.

The film composition control can be translated into the volatile element (e.g. Lead) control within film. Various factors that influence the volatile element within film are thought to exist (Fig. 1) and we have investigated the influence of sputtering conditions, strength of magnetic field and electric potential of substrate.

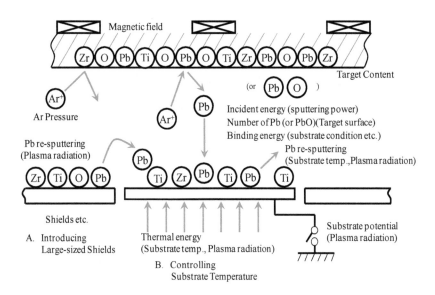

Figure 1. Various factors that influence content with PZT Film.

In addition to the known characteristics in perovskite oxide thin-film sputtering that volatile elements easily fluctuate, it was confirmed that volatile elements within film were unstable even when the deposition was performed under identical conditions. The conceivable causes for that phenomenon were the instability of the target, fluctuation of temperature in the sputtering chamber, and variation of plasma status over time.

Main problem in perovskite oxide thin-film is the change in the volatile element content with the passage of sputtering time. As for the PZT, continuous sputtering of 2.0 kWh showed approximately a 30% decline in lead content compared to that at the beginning. It was determined that the reason for this problem was the change in plasma status. As shown in Fig. 2, when insulating PZT film adhered to the shields of the ground potential, charge-up occurred, the impedance of the system changed, plasma was pushed to the center of the chamber, exposure to plasma was enhanced, and as a consequence, Pb content within film is reduced. In order to stabilize the status of plasma, we installed a stable anode, that is, an anode that avoided charge-up due to the adhesion of insulating PZT film and maintained the role as an anode. Consequently, as can be seen in Fig. 3, stability of Pb content within film in continuous sputtering has been confirmed.

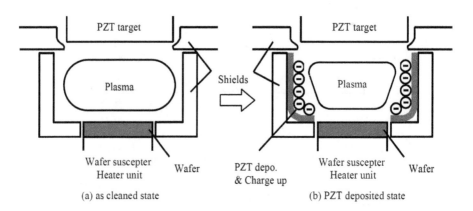

Figure 2. Change in plasma status which is responsible for Pb content variation.

Fig. 4 shows an example of thickness and Pb content uniformity in a PZT film on 8″ substrate. Both thickness and Pb content uniformity varied according to the deposition condition and was minimized around the sputtering pressure of 1.0 Pa. Thickness and Pb content uniformity represent good result as low as ±1.9% and ±1.1%, respectively. These uniformities were also confirmed as stable, thereby satisfactorily meeting the requirements of mass production.

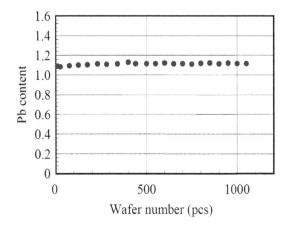

Figure 3. Stable transition of Pb content within film in continuous sputtering.

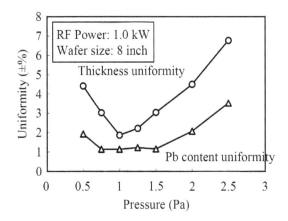

Figure 4. Uniformity in 8 inch area.

2.3. Sputtered PZT thin-films for non-volatile memory application

In this section, we introduce one of its achievements, the ferroelectric characteristics of PLZT capacitors for FeRAM. We used the multi chamber type mass production sputtering system equipped with an exclusive sputtering module for ferroelectric materials, CERAUS ZX-1000 from ULVAC. This system has the following features in addition to features such as easy maintenance, short exhausting time, short down time, etc.

1. Can mount 300 mm (12 inch) in diameter targets and process large diameter substrates of 200 mm. At present, deposition of PZT thin-films on 6 and 8 inch substrates are performed using 12 inch single ceramic target.

2. Including the heat chamber, this system has five process chambers, thereby achieving high flexibility. The system is presently executing the following in-situ processes as standard: pre-heating of substrate → substrate sputtering (e.g. Ti, TiN, Pt) → ferroelectric material sputtering.

3. As a substrate heating mechanism, this system was capable of precise and rapid heating in a wide range from low to high temperatures, with the aid of an electrostatic chuck type hot plate, in addition to lamp heating.

4. This system used RF sputtering for ferroelectric deposition and counters RF noises.

Fig. 5 shows the transition of switching charge (Q_{SW}) and saturation characteristics of a Pt/PZT(200 nm)/Pt capacitor measured at 5 V. Q_{SW} with 5 V applied was approximately 34 $\mu C/cm^2$ and saturation voltage of 90% ($V_{90\%}$) was 3.1-3.2 V. The composition of PLZT film, which contained added Ca and Sr for the improvement of retention and imprint characteristics, was excellent.

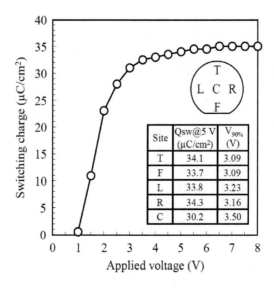

Figure 5. Transition of switching charge of Pt/PZT(200 nm)/Pt capacitor.

Fig. 6 shows the stability of Q_{SW} in continuous sputtering of 1000 substrates. The result showed high stability sufficient for mass production. Also, reference data was the data before chamber cleaning. The fact that it was equivalent to the data after chamber cleaning demonstrates the reliability of the system process.

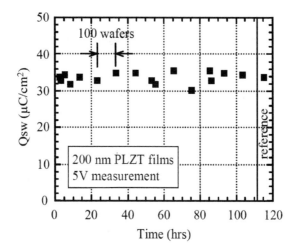

Figure 6. Stability of Qsw in continuous sputtering of 1000 substrates.

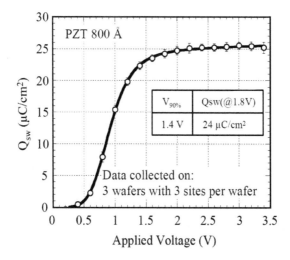

Figure 7. Transition of switching charge of IrOx/PZT(80 nm)/Pt capacitor.

Further improvement of ferroelectric performancs is needed because scaled thinner capacitor (sub 100 nm) is demanded for next generation. Pb deficient surface layer is confirmed to be responsible for degradation of ferroelectric performance. Bottom electrode and PZT deposition process were modified to further improve the ferroelectric performance and achieve thinner capacitor with good performances. [8]. Fig. 7 shows the Q_{SW} transition curve of a Ir-Ox/PZT(80 nm)/Pt capacitor measured at 3 V. Q_{SW} with 1.8 V applied was approximately 24 $\mu C/cm^2$ and $V_{90\%}$ was 1.4 V. Fig. 8 also shows the fatigue characteristics of voltage application at 2 V. Q_{SW} was not decreased even after switching in 10^9 cycles.

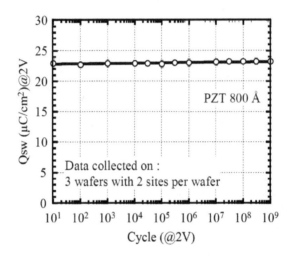

Figure 8. Fatigue characteristics of IrOx/PZT(80 nm)/Pt capacitor.

From above results, sputter-derived PZT capacitor was proved to be suitable for 0.18 μm technology node.

2.4. MOCVD technology for non-volatile memory application

For next-generation FeRAMs beyond 0.18 μm, because it is necessary to achieve further larger packing densities and integration with logic devices, thinner films of ferroelectric associated with shrinking of the thickness of ferroelectric capacitors and the three-dimensional structures associated with shrinking of the capacitor areas are demanded. In addition, it is necessary to achieve thinner films of ferroelectric in parallel with its higher quality in order to meet the demand of still lower voltage drives for device performance. It is said that MOCVD (Metal organic chemical vapor deposition) technology can meet these demands. While depositions are physically made by plasma collision in sputtering processes, depositions are made on heated substrates by chemical reaction among source gases in MOCVD. Therefore, denser crystalline films are easily obtainable, and it is possible to achieve thinner films and higher quality. In addition, uniform deposition can be obtained also on three-dimensional structure, and it is considered that good step coverage can be obtained. Because of these features, MOCVD is the prime candidate of deposition technologies for next-generation FeRAMs in mass production, and its development is making progress.

First, the features of MOCVD are briefly explained. In the method of MOCVD, raw material that was changed into organic metal with high vapor pressure is led to a substrate, and is thermally reacted with a reactant gas (such as oxygen) on the substrate to form a deposition. Methods for the gasification of organic metals are classified into two groups. One of them is the sublimation of solid raw materials (sublimation method). After a certain vapor pressure is obtained by heating a solid raw material using a heater, it is transported with carrier gas. In the second method, a liquid raw material or solid raw material is melted in an organic solvent. Then, after the solution is vaporized, it is transported with carrier gas (vaporization method). In the case of the former, the deposition rate is small. In the case of the latter, there is a problem of the instability of vaporization. For the target of mass-production system development, we adopted the latter vaporization method from the beginning of the development. The problems that need to be resolved are as follows (see Fig. 9):

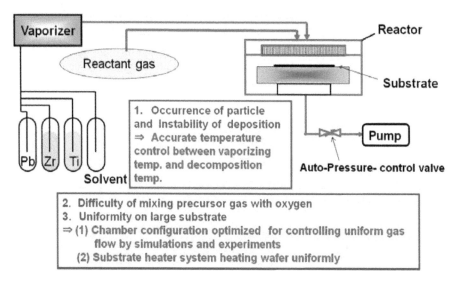

Figure 9. Features of MOCVD Module

1. Vaporization and transport of multi-element raw materials In MOCVD using the vaporization of a ferroelectric solution, because a multi-element organic metal material is vaporized and transported, stable vaporization is difficult, and there is a possibility that precipitation or decomposition may occur in a pipe during transport.

2. Mixture of gases whose molecular weights are mutually different It is difficult to uniformly mix a reactant gas such as oxygen and an organic raw material whose molecular weights are vastly different.

3. Uniformity of large-diameter substrates For the film thickness and the composition of a multi-element oxide thin film PZT, in-plane uniformity in 8 inches is required.

4. PTZ film properties Because of high-temperature processes and in-situ crystallization in the processes, the correlation between the kind of a raw material and a process is sensitive, and the control of a film composition and crystalline orientation is difficult. Therefore, necessary performance of films is difficult to obtain.

In order to establish MOCVD mass-production technology, it is absolutely necessary to resolve the above problems in parallel, and advanced vaporization technology, mixing technology, and reaction control technology are simultaneously required. We combined ferroelectric deposition technology for FeRAMs, module design technology, and CVD equipment technology, which were provided to users until now, and completed full-fledged MOCVD equipment for mass production. The major features include the following four points:

1. The reproducibility in the continuous operation of vaporization and the gas transport to a substrate that controls condensation and decomposition were achieved by accurate temperature control for each part of equipment and the optimization of vaporization conditions. Consequently, film thicknesses within ±2% and the reproducibility of the PZT composition were kept with no mechanical maintenance, and a running test for 1,000 substrates was successful as shown in Fig. 10. [9]

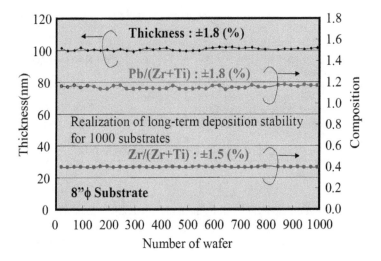

Figure 10. Reproducibility in MOCVD system (film thickness and PZT composition)

2. Because the design of the equipment was conducted in consideration of the flow of gases and the mixture between a raw material gas and a reactant gas on the basis of simulations, film thickness distribution within ±3% on 8 inch substrates was achieved as shown in Fig. 11.

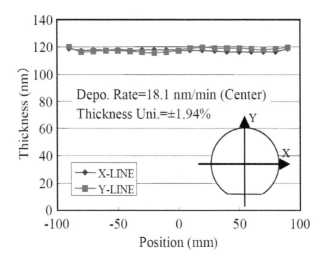

Figure 11. In-plane uniformity of PZT film thickness

3. Because of the development of a new heater, including the optimization of the shape of the heater, the temperature distribution of 8 inch substrates can be continually controlled within ±3ºC. Consequently, in-plane uniformity within ±3% was achieved for both film thicknesses and the composition as shown in Fig. 12. [10]

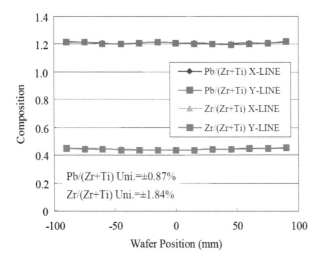

Figure 12. In-plane uniformity of PZT composition

4. By taking advantage of the progress of the equipment hardware as described above, the optimization of processes was conducted, and PZT thin films can be controlled to ensure preferential orientation in the <111> direction as shown in Fig. 13. Consequently, the formation of PZT films within 100 nm that have capacitor properties with a 1.5 V low voltage drive is achieved as shown in Fig. 14. [11,12] In addition excellent endurance properties which are over 10^{10} cycles were obtained for 73nm-PZT in Fig 15.

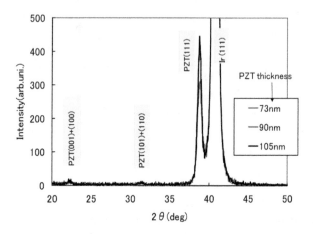

Figure 13. XRD spectrum for various thickness PZT

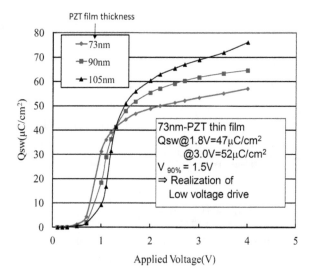

Figure 14. Characteristics of MOCVD-PZT film (applied voltage dependence of switching charge)

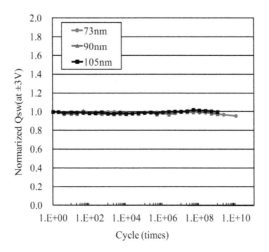

Figure 15. Endurance properties of MOCVD-PZT films

Recently mass production tool for 300mm Si wafer was developed and excellent in-plane uniformity less than 1.5% for thickness and PZT composition were obtained.

2.5. Etching technology

2.5.1. Issues of ferroelectric etching technology

Conventionally, the piezoelectric elements have been fabricated by chemical wet etching [13] or argon ion milling. With the miniaturization of MEMS, there have been increasing demands for dry etching with the excellent shape controllability as semiconductor technology. Recently dry etching technique for MEMS using PZT was reported. [14,15] The Pt, Ir and other rare metal electrodes and the PZT ferroelectric thin films that compose piezoelectric elements react poorly with halogen gases and their halides have low vapor pressures. For these reasons, these materials are called hard-to-etch materials. The following technical issues are important for dry etching of the PZT ferroelectric thin films for not only FeRAM but also MEMS productions:

1.   Etching selectivity to resist mask and the bottom rare metal electrode

A piezoelectric element film consists of PZT with a thickness of several micrometers and the rare metal electrodes with a thickness of about 100 nm. Generally, the bottom electrode is left after the PZT etching. Therefore, a low etching rate for the bottom electrode, the so-called high etching selectivity, is important as a PZT etching condition.

2.   Adhesion of conductive deposit to the pattern sidewalls and damage to PZT

The materials are hard to etch, and their etching products easily adhere to the pattern sidewalls, and result in leaks between the top and bottom electrodes. What is worse, the pattern sidewalls are exposed to reactive gas plasma during etching, and tend to suffer lead and oxygen coming out and other damages.

3.   Plasma stability during continuous processing

Adhesion of etching products to chamber walls, especially the RF introduction window that generates plasma, causes instabilities of plasma and deteriorates the etching rate and the shape reproducibility. Avoiding of adhesion of etching products to chamber walls is important for mass production.

4.   Uniformity of etching rate within wafer

As in the case of (1), to stop the thick PZT at the thin bottom electrode after etching, the uniformity of etching rate within wafer is important.

2.5.2. Ferroelectric etching systems and process for mass production

As the piezoelectric PZT etching systems, this section explains about Apios NE series made by ULVAC, Inc. The etching module is equipped with the ISM (Inductively Super Magnetron) plasma source that can generate low-pressure and high-density plasma. Fig. 16 shows a drawing of the etching module. Table I shows the comparison between the normal ICP type plasma source and the ISM plasma source. The RF antenna is mounted in the upper part of the etching chamber, so that RF is introduced through the quartz window into the etching chamber to generate plasma. The uniformity of the etching rate within wafer can be easily optimized by positioning permanent magnets under the antenna. Fig. 17 shows the uniformity of PZT etching rate within 6 inch wafer. A high uniformity (<+/-5%) was realized by means of optimization to permanent magnet layout. A STAR electrode is provided between the antenna and the quartz window to control adhesion of etching products to the quartz window by applying RF to the STAR electrode. The substrate is held on the electrostatic chuck. The substrate temperature is controlled by introducing Helium to the backside of the substrate. The ion energy is controlled by applying RF power to the substrate. The materials to be etched are non-volatile, and the etching products adhere to the shield located in the chamber. The temperature of the shield is kept constant by heater, so process is high stability.

	ISM	ICP
Plasma density (cm^{-3})	$1\times10^{10} \sim 1\times10^{11}$	$5\times10^{9} \sim 5\times10^{10}$
Operating pressure (Pa)	0.07<P<7	0.5<P<50
Uniformity	Optimized magnetic layout	Determined by chamber structure
Damage	Plasma density and substrate bias can be controlled independently.	
Repeatability, stability	Less re-deposition results in Low-pressure etching -> better repeatability	High pressure process causes re-deposition.
Maintenance	Chamber structure is simple for easy maintenance.	

Table 1. Inductively super magnetron (ISM) plasma performances.

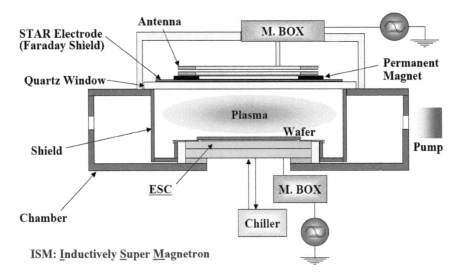

Figure 16. Etching chamber with inductively super magnetron (ISM) plasma performances

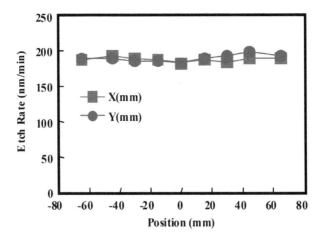

Figure 17. PZT etching uniformity

2.5.3. Mass productive high temperature etching technology for high dense FeRAM

Dry etching techniques are used for the patterning of FeRAM device. It is difficult to etch the material of FeRAM such as the noble metal and PZT, because these have low reactivity with halogen gas plasma and these halides have low vapor pressure. The photo-resist is used as etching mask for the patterning of FeRAM memory cell. But the etching selectivity to photo-resist is low. Therefore the etching profile becomes low taper angle. Since FeRAM device shrinks down recently(0.35-μm-design rule or lower), high temperature etching was developed for high density FeRAM device. Furthermore, FeRAM device changed to the stack structure from planer structure. At that time, the top electrode, PZT and the bottom electrode are etched in a series by using of hard-mask (Fig. 18). [16]

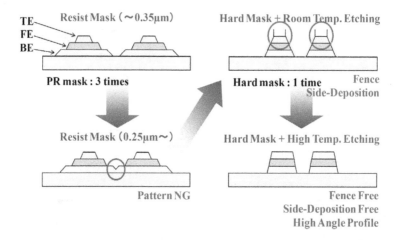

Figure 18. Necessity of high-temperature etch

This section explains high temperature etching system ULHITE series made by ULVAC, Inc. This etching module is equipped with ISM plasma source. Most important feature is the novel electro-static-chuck (ESC) type hot plate stage at a temperature up to 450°C, and this stage can be supplied high bias power. The process chamber, variable conductance valve, pumps and gas exhaust are heated up. The deposition shields equipped in process chamber are also heated up to 200°C. The effect is reducing the deposition during the etch process, and high process stability is achieved.

Fig. 19 shows repeatability of Ir etching time for 300-wafers running test. The plasma cleaning were carried out in every 25-wafers. Excellent stability was confirmed. For next generation FeRAM devices, the continuous etching among the top electrode, ferroelectrics and bottom electrode by use of one mask is need. And the high taper angle and no sidewall deposition are needed for etching process.

Fig. 20 shows the results of FeRAM capacitor etching of stack structure dependence of the stage temperature. The halogen gases were used in etching process. Thus the higher profile can be obtained by higher stage temperature. Fig. 21 shows repeatability of etching profile for 300-wafers running test. Excellent repeatability of etching profile was confirmed.

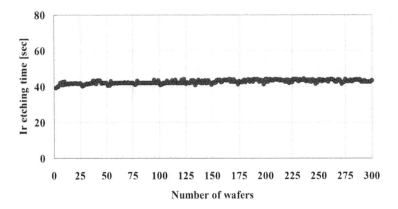

Figure 19. Repeatability of Ir etching time for 300-wafers running test

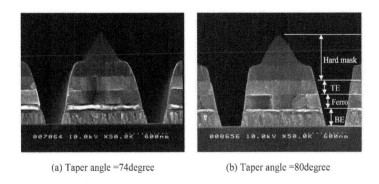

(a) Taper angle =74degree   (b) Taper angle =80degree

Figure 20. Stage temperature dependency of taper angle ((a) 300°C, (b) 400°C)

Usually, high density device is required for non-volatile memory. Therefore the high temperature etching technique contributes to realize the next generation FeRAM devices production.

3. Technology for MEMS application

3.1. Sputtered PZT piezoelectric films for MEMS application

A multi-chamber type mass production sputtering systems for electronic devices SME-200 equipped with an exclusive sputtering module described above. PZT films have been deposited on 6 inch diameter silicon substrates.

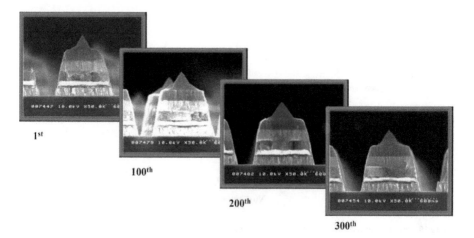

Figure 21. Repeatability of etching profile for 300-wafers running test

The platinum bottom electrodes, whose orientation is (111), were deposited on the substrate. The PZT films were deposited under Ar/O$_2$ mixed gas atmosphere of 0.5 Pa. Substrate temperature was heated up to around 550°C. After the deposition, PZT films were conducted with no thermal treatments such as post-annealing. PZT films were deposited with relatively high growth rate about 2.1 µm/h and these thicknesses were from 0.5 to 3.0 µm in consideration of piezoelectric MEMS applications. The ceramic target with Zr/Ti ratio of 52/48, in which 30 mol % excess PbO was added for the compensation of the lead re-evaporation from the films, was used in order to obtain PZT films near the stoichiometric composition. After the PZT deposition, top electrode 100-nm-thick Pt was deposited by the dc sputtering method.

For the measurement of the piezoelectric properties of the PZT films, Rectangular beams (cantilevers) with the size of about 30 mm (3 cm) × 3 mm were prepared. Polarization and displacement in these films were simultaneously observed using the laser doppler vibrometer (Graphtec AT-3600) and the laser interferometer (Graphtec AT-0023) which were attached to a ferroelectric test system.

Fig. 22 shows relationship between Pb composition, deposition rate and repeatability. As a result, stable transition of Pb content within film in continuous sputtering has been confirmed. As can be seen from the figure, the change in the deposition rate was 2.1 µm±1.4%, and changing Pb composition was 1.0±0.1% in the short running for 35 pieces (total thickness; 105 µm).

Fig. 23 shows XRD patterns of as-sputtered PZT film. No pyrochlore phase can be confirmed and the film appears to be almost perovskite phase with the preferred orientation to (001) or (100). It confirmed that the uniformity of the crystalline property were good in 6 inch area. In addition, the dielectric constant of these samples was measured as shown in Fig.24. As a result, excellent properties with the dielectric constant over 1000 and its uniformity of ±4.7% were confirmed.

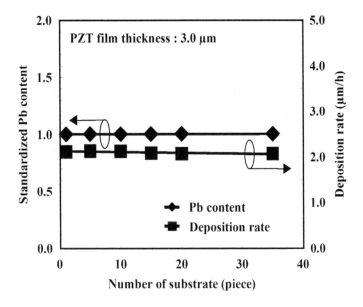

Figure 22. Relationship between Pb composition, deposition rate and repeatability

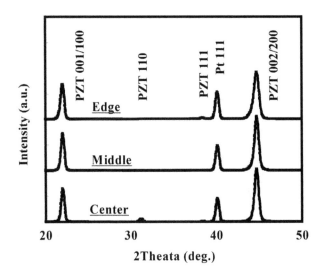

Figure 23. Crystallization uniformity of PZT films Deposited on 6-inches substrate

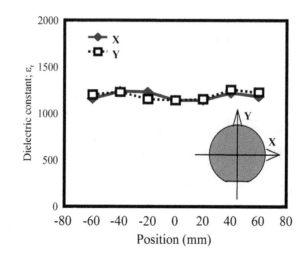

Figure 24. Dielectric constant of PZT films deposited on 6 inch substrate

Piezoelectric properties were finally confirmed for PZT films by checking the cantilever as shown in Fig. 25. Large piezoelectric coefficient from -60 to -120 pm/V was observed in our PZT films.

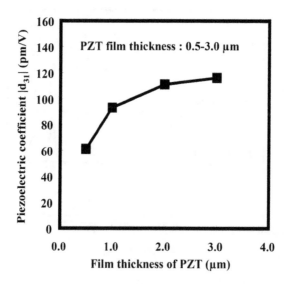

Figure 25. Relationship between piezoelectric coefficient and PZT film thickness

3.2. Etching technology of PZT piezoelectric Films for MEMS application

Additionally etching technology for mass production is developed. Since high etching rate is required in MEMS PZT process because of large PZT thickness, etching selectivity to the bottom electrode(Pt) and uniformity are important. In Fig. 26 we show etching rate profile of both electrode(Pt) and PZT film. It has been found that etching rate of Pt is smaller than PZT over whole wafers, implying that Pt film is the role of an etch stop layer.

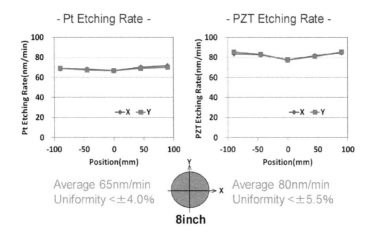

Figure 26. Etching rate and in-plane uniformity of Pt & PZT

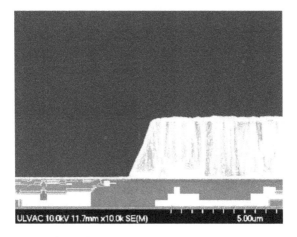

Figure 27. SEM image of Pt/PZT etching profile

Fig. 27 shows the SEM image after the piezoelectric element was etched and the resist mask was removed. The film composition is Pt/PZT/Pt=100 nm/3 μm/100 nm, and the top Pt electrode and PZT were continuously etched by using a 5-μm-thick photo resist as a mask. Chlorine and fluorine mixed gases are used for PZT etching. The etching shape (taper angle) is about 65º, and nothing adhered to the pattern sidewalls. Despite 20% of over etching, the bottom Pt electrode was hardly etched. This indicates that a high etching selectivity to Pt was achieved. Fig. 28 shows the dependency of the etching rate and the taper angle on the bias power. As the bias power increases, the etching rate increases linearly. When 400 W was applied, the PZT etching rate of 190 nm/min. was achieved. The taper angle also increased gradually as the bias power increased.

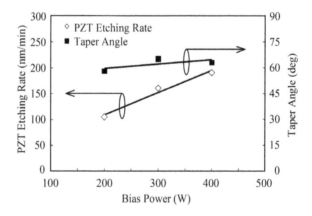

Figure 28. Dependence of PZT etching rate and taper angle on the substrate bias power

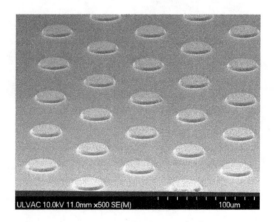

Figure 29. Pt/PZT etching profile of φ50mm piezoelectric element array

Fig. 29 shows a 50-μm-diameter actuator element array fabricated by dry etching. Fig. 30 shows that the remanent polarization (Pr value) of the piezoelectric thin-film actuator with a 3-μm-thick PZT film was 40.5-42.8 μC/cm² and the coercive electric field (Eᶜ value) was 44.5-46.0 kV/cm at an applied voltage of 30 V, and the characteristics without the dependency on element size (30-300 μm diameter) were obtained. The damage to the PZT piezoelectric thin-film actuator caused by the dry etching is considered to be negligible. The displacement of the PZT thin-film actuator was measured by contact-AFM. Fig.31 shows that a displacement of about 4 nm was obtained at 3-μm-thick PZT film, 30-μm-diameter element size, and an applied electric field of 100 kV/cm. It was clarified that the processing of PZT piezoelectric thin-film actuators by dry etching is very effective.

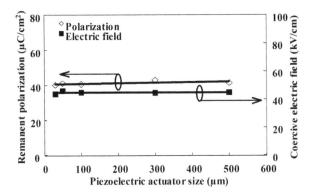

Figure 30. Dependence of ferroelectric properties on piezoelectric actuator size

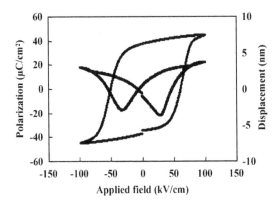

Figure 31. Ferroelectric and displacement properties of Pt/PZT/Pt element with 30mm diameter etching rate and taper angle on the substrate bias power

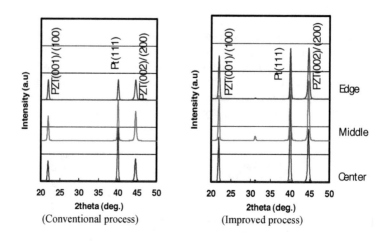

Figure 32. XRD Patterns of PZT film before and after improvement on 8inch Pt/Ti/SiO$_2$/Si Substrate

3.3. Further development for MEMS application

We also note that post in-situ treatment [17] has been recently established, aiming to improve crystalline property and piezoelectric coefficient in our PZT films. Fig. 32 shows the results before and after improvement of crystalline properties. Peak intensity of a/c axis of PZT film deposited by improved process is approximately two times higher of peak intensity of PZT film deposited by conventional process uniformity over 8 inch wafers. As a consequence good piezoelectric coefficient (12.9C/m^2) and large breakdown voltage (68V) were obtained for PZT film deposited by improved sputtering process in Fig 33.

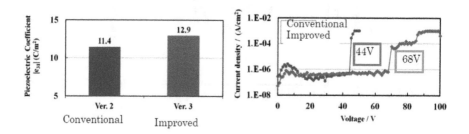

Figure 33. Improvement of piezoelectric coefficient(Left) and break down voltage(Right)

Good piezoelectric performance which is based on mass production technology including excellent in-plane uniformity and process rate of sputtering and etching was obtained in above evaluation.

3.4. Sputtered BST thin-films for capacitor and RF tunable devices

(Ba,Sr)TiO$_3$ (BST) is expected to use as the thin film capacitor, RF tunable component [18, 19] on-chip capacitor [20] for its excellent dielectric behavior. In these applications, decoupling, high permittivity and high tunability are required as dielectric characteristics. High permittivity is thought to be primarily important character as dielectric characteristics. BST thin films have been deposited by some deposition techniques. Among these techniques, RF magnetron sputtering is thought to be suitable for mass-production because of its stability and reproducibility as well as good film performances. In this section, BST films deposition by RF magnetron sputtering and an approach to deposit BST films with higher permittivity were described.

BST films were also deposited by an RF magnetron sputtering method using sintered BST ceramic targets. Basic sputtering conditions such as RF power, deposition temperature and so on were varied. As a top electrode, Pt dot with 0.5 mm in diameter was deposited by DC magnetron sputtering. After top electrode deposition, BST capacitors were post-annealed for 1 hour in oxygen atmosphere under 1 atm using resistive heat furnace.

Fundamental properties of BST thin films such as dielectric constant, crystalline quality (XRD), surface morphology (SEM) were investigated. Agilent 4284A LCR meter was used for dielectric properties measurement and measuring cycle and volts alternating current were 1 kHz and 1 Vrms, respectively.

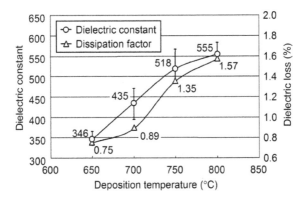

Figure 34. Relationship between dielectric constant and deposition temperature

The relationship between dielectric constant and deposition temperature was investigated as shown in Fig. 34. Ba/Sr ratio of BST target was 50/50. RF power was 1500 W. Ar/O$_2$ flow ratio was 3. Sputtering pressure was 2.0 Pa. BST film thickness was 100 nm. As a result, deposition rate was 4.8 nm/min and almost constant irrespective of deposition temperature. As can be seen in this figure, dielectric constant is strongly dependent on deposition temperature and increasing with deposition temperature increasing. XRD patterns of these films are also shown in Fig. 35. We can see in this figure, BST grains are randomly orientated and XRD peak intensities from BST films are increasing with deposition temperature increasing. So, it is thought that this represents the relationship between dielectric constant and deposition temperature.

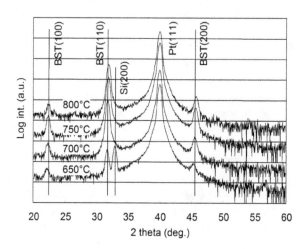

Figure 35. Relationship between XRD pattern and deposition temperature

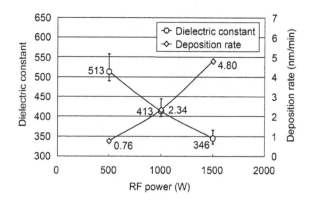

Figure 36. Relationship between dielectric constant, deposition rate and RF power temperature

The relationship between dielectric constant, deposition rate and RF power is shown in Fig. 36. Deposition temperature was 650°C in this relationship, while other conditions were not changed. We can see in this figure, there is trade-off relationship between dielectric constant and deposition rate and higher dielectric constant can be obtained at low sputtering power of 500 W in this experiment. BST morphology (SEM photographs) dependence on RF power is also shown in Fig. 37. As can be seen in this figure, BST film deposited at 500 W (b) has dense and uniform columnar structure. Possibly it represents some relationship between dielectric constant and RF power. It is speculated that there is a long time for atomic migration and the growth of BST crystal is encouraged under such low deposition rate around 1 nm/min.

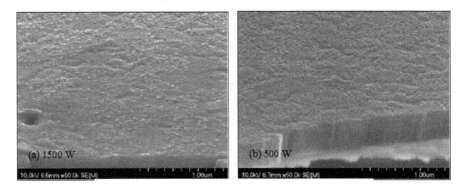

Figure 37. BST morphology dependence on RF power

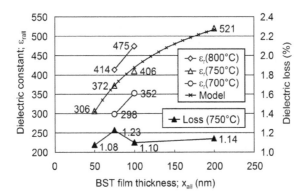

Figure 38. Relationship between dielectric constant and BST film thickness

Furthermore, the relationship between dielectric constant and BST film thickness is shown in Fig. 38. Ba/Sr ratio of BST target was 70/30 in this experiment. Deposition temperature was varied from 700°C to 800°C as a parameter, while other conditions were not changed.

We can see in this figure, dielectric constant is also strongly and non-linearly dependent on BST film thickness.

If BST capacitor is assumed to be a simple series connected capacitor between transition layer near BST/Pt bottom electrode interface and main layer which represents BST film excluding transition layer, it is speculated that there have been the transition layer in this BST film. [21]

As a simple experiment, effects of gas flow sequence for BST deposition was investigated. Gas flow sequence, that is ON/OFF step was changed, while Ar/O$_2$ flow ratio was not changed. Dielectric constant was improved by introducing of oxygen gas before BST deposition. It is speculated in this experiment that BST/Pt interface was improved because the oxygen vacancies of BST in this region were reduced. Therefore, dielectric properties are noticeably influenced by gas flow sequencing variation.

4. Technology for ferroelectric probe memory

Technology for ferroelectric probe memory was developed with the deposition technology for FeRAM manufacturing. Hard-disk using the magnetic recording is the one of the major storage device, but the recording density will reach the limit in the near future by exteriorization of the magnetization disappearance by the heat disturbance. Ferroelectric probe memory is ultrahigh-density memory applied Scanning Probe Microscope (SPM) and ferroelectric property.

Ferroelectric perovskite Pb(Zr, Ti)O$_3$ (PZT) is polarized by electric field. For scanning probe memory device, ferroelectric layer required atomically-smooth surface and low leakage current to achieve large recording density by forming polarized domains as unit cell size. Therefore it is necessary to grow the epitaxial ferroelectric thin films oriented to c-axis direction.

In this time we prepared epitaxial growth PZT/ SrRuO$_3$ (SRO) thin films on single crystal SrTiO$_3$ (STO) substrate for probe memory device, and evaluated the film properties and the recording density with Piezoresponse force microscopy (PFM). The SRO films were deposited on STO (100) single crystal substrate with DC sputtering. After that, the PZT film was deposited by MOCVD process.

Fig. 39 shows the AFM image of PZT / SRO / STO deposited at the optimal Pb flux. The surface morphology of PZT has step structure and smooth terraces similar to SRO/STO substrate before PZT deposition. This surface profile indicates layer-by-layer growth of PZT on SRO / STO substrate [22] [23] [24]. Fig. 40 shows XRD spectrum of PZT deposited at Pb flux of 0.160ml/min. There are only PZT, SRO and STO peaks. The fringe pattern appears around PZT (001) and (002) peaks because of the smooth interface between PZT and SRO. Full-width at half-maximum (FWHM) of PZT (002) with X-ray rocking curve is 0.129degree for optimal Pb flux of 0.160ml/min. Rms of surface roughness and PZT(002) FWHM of X-ray rocking curve have a similar trend relative to Pb flux.

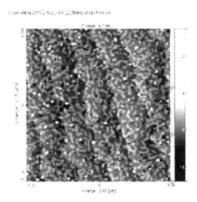

Figure 39. AFM image of PZT at Pb flux of 0.160ml/min (Scanning range is 2um square)

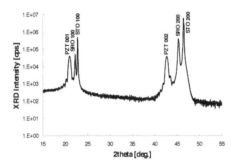

Figure 40. XRD spectrum of PZT / SRO /STO

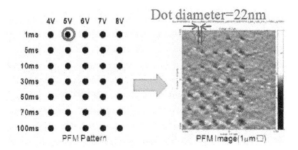

Figure 41. Writing Voltage & Writing Time Dependency of Dot diameter – [Small diameter (22nm) of recording-dot was obtained for 5V & 1ms writing condition. (PZT/SRO/c-STO) ・ Estimated Recording density: 1.4Tbit/inch².]

PFM measurement is carried out with various biases between 4V and 8V and pulse widths between 1msec and 100msec with $PtIr_5$ coatedprobe (tip diameter is 25nm). Fig. 41 shows PFM write-condition and PFM image of the result. Polarization-inverted dot diameter has a minimum of 20nm at the bias of 5V and pulse width of 1msec. Recording density is estimated on the assumption that this tiny dot is arrayed with pitch of 20nm. From the calculation result, device sample deposited under the optimal growth condition achieves large bit density of $1.4Tb/inch^2$.

5. Conclusion

We have been developing thin film process technologies for ferroelectric application of advanced semiconductor and electronics usage for 20 years or more, and completed the ferroelectric thin film solutions (sputtering, MOCVD, and etching) that became a de facto standard. These technologies will support a wide variety of convenient energy-saving devices such as FeRAM, MEMS production (actuators composing gyro meters, portable camera modules for smart phone applications, tunable devices and so on), and ultra-high density probe memory.

Author details

Koukou Suu

Institute of Semiconductor and Electronics Technologies, ULVAC, Inc., Shizuoka, Japan

References

[1] K. Suu, A. Osawa, N. Tani, M. Ishikawa, K. Nakamura, T. Ozawa, K. Sameshima, A. Kamisawa and H. Takasu: Jpn. J. Appl. Phys. 35 (1996) 4967.

[2] K. Suu, A. Osawa, N. Tani, M. Ishikawa, K. Nakamura, T. Ozawa, K. Sameshima, A. Kamisawa and H. Takasu: Integr. Ferroelectr. 14 (1997) 59.

[3] K. Suu, A. Osawa, Y. Nishioka and N. Tani: Jpn. J. Appl. Phys. 36 (1997) 5789.

[4] K. Suu, Y. Nishioka, A. Osawa and N. Tani: Oyo Buturi 65 (1996) 1248. (in Japanese)

[5] K. Suu: Proc. Semicon Korea Tech. Symp., 1998 p. 255.

[6] N. Inoue, Y. Maejima and Y. Hayashi: Int. Electron Device Meet. Tech. Dig., 1997 p. 605.

[7] N. Inoue, T. Takeuchi and Y. Hayashi: Int. Electron Device Meet. Tech. Dig., 1998 p. 819.

[8] F. Chu, G. Fox, T. Davenport, Y. Miyaguchi and K. Suu: Integr. Ferroelectr. 48 (2002) 161.

[9] T. Yamada, T. Masuda, M. Kajinuma, H. Uchida, M. Uematsu, K. Suu and M. Ishikawa : IFFF2002 abstract, 4 (2002) 37.

[10] T. Masuda, M. Kajinuma, T. Yamada, H. Uchida, M.Uematsu, K. Suu and M. Ishikawa : Integr. Ferroelectr., 46 (2002) 66

[11] Y. Nishioka, T. Jinbo, T. Yamada, T. Masuda, M. Kajinuma, M. Uematsu, K. Suu and M. Ishikawa: Integr. Ferroelectr., 59 (2003) 1445

[12] Y. Nishioka, T. Masuda, M. Kajinuma, T. Yamada, M. Uematsu and K. Suu: MRS Fall Meeting Proceedings, 784-C7 (2003) 6

[13] K. Zheng, J. Lu, and J. Chu: Jpn. J. Appl. Phys. 43 (2004) 3934.

[14] Y. Kokaze, M. Endo, M. Ueda, and K. Suu: 17th Int. Symp. Integrated Ferroelectrics, 2005, 5-26-P.

[15] Y. Kokaze, I Kimura, M. Endo, M. Ueda, S. Kikuchi, Y. Nishioka, and K. Suu: Jpn. J. Appl. Phys. 46 (2007) 282.

[16] M. Endo, M. Ueda, Y. Kokaze, M. Ozawa, T. Nakamura, K. Suu: SEMI Technology Symposium 2002; December 5, 2002

[17] Suu et.al in preparation

[18] I. P. Koutsaroff, T. Bernacki, M. Zelner, A. Cervin-Layry, A. Kassam, P. Woo, L. Woodward, A. Patel, Mat. Res. Soc. Symp. Proc. Vol. 762, C5.8.1 (2003).

[19] X. H. Zhu, J. M. Zhu, S. H. Zhou, Z. G. Liu, N. B. Ming, S. G. Lu, H. L. W. Chen, C. L. Choy, Journal of Electronic Materials, 32 (2003) 1125.

[20] M. C. Werner, I. Banerjee, P. C. McIntyre, N. Tani, M. Tanimura, Appl. Phys. Lett., 77 (2000) 1209.

[21] T. Jimbo, I. Kimura, Y. Nishioka and K. Suu, Mat. Res. Soc. Symp. Proc. Vol. 784, C7.8.1 (2003).

[22] P.-E. Janolin, B. Fraisse, F. Le Marrec, and B. Dkhil, Appl. Phys. Lett. 90, 212904 (2007)

[23] Keisuke Saito, Toshiyuki Kurosawa, Takao Akai, Shintaro Yokoyama, Hitoshi Morioka, Takahiro Oikawa, and Hiroshi Funakubo, Mater. Res. Soc. Symp. Proc. 748, U13.4.1-U13.4.6 (2003).

[24] H. Hu, C. J. Peng and S. B. Krupanidhi, Thin Solid Films, 223 (1993) 327 333I. G. Baek, et al,: Tech. Dig. Int. Electron Devices Meet., San Francisco, 2004, 23, 6, p58

Chapter 6

Ferroelectrics at the Nanoscale: A First Principle Approach

Matías Núñez

Additional information is available at the end of the chapter

1. Introduction

The field of ferroelectric materials is driven by its possible use in various micro-electronic devices that take advantage of their multifunctional properties. The existence of a switchable spontaneous polarization (see Figure 1) is at the basis of the design of non volatile ferroelectric random access memories (FERAMs), where one bit of information can be stored by assigning one value of the Boolean algebra ("1" or "0") to each of the polarization states. Also, the high dielectric permitivity of ferroelectrics makes them possible candidates to replace silica as the gate dielectric in metal-oxide-semiconductor field effect transistors (MOSFETs). Their piezoelectric behavior enables them to convert mechanical energy in electrical energy and vice versa. Their pyroelectric properties are the basis for highly sensitive infrared room temperature detectors.

The current miniaturization of microelectronic technology, imposed by the semiconductor industry, raises the question of possible size effects on the properties of the components. Except for the case of the gate dielectric problem in MOSFET's, the thickness of the films used in contemporary applications is still far away from the thickness range where size effects become a concern; therefore the question at the moment is merely academic. Nevertheless, it is very possible that the fundamental limits of materials might be reached in the future.

Despite the efforts and advancement in the field, both theoretically and experimentally (see a recent review in Ref.[1]), many questions still remain open. The main reason for the poor understanding of some of the size effects on ferroelectricity is the vast amount of different effects that compete and might modify the delicate balance between long range dipole-dipole electrostatic interactions and the short range forces, whose subtle equilibrium is known to be at the origin of the ferroelectric instability.

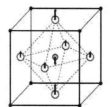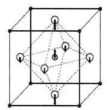

Figure 1. Tetragonal ferroelectric structure of $BaTiO_3$. Solid, shaded and empty circles represent Ba, Ti, O atoms, respectively. The arrows indicate the atomic displacements, exaggerated for clarity. The origin has been kept at the Ba site. Two structures, with polarization along [001], are shown. Application of a sufficiently large electric field causes the system to switch between the two states, reversing the polarization. in principle, we can assign one value of the Boolean algebra ("1" or "0") to each of the polarization states.

The next section will give an overview of the main concepts involved in the modern theory of polarization. A definition of a local polarization will be given in terms of the centers of the wannier functions associated with the band structure of the system. Next, some basic electrostatic notions related with ferroelectric films will be given, in particular the concept of depolarization field and screening by metal contacts. Finally, the results of this research work: applying the layer polarization concept, where we show the hidden structure of the polarization at the nanoscale.

2. Modern theory of polarization

Classically, the macroscopic polarization in dielectric media is defined to be an intensive quantity that quantifies the electric dipole moment per unit volume [2–4]. Definitions along these lines work well for finite systems but have important conceptual problems when applied to periodic crystalline systems because there is no unique choice of cell boundaries [5].

In the early 1990s a new viewpoint emerged and lead to the development of a microscopic theory [6–8]. It starts by recognizing that the bulk macroscopic polarization cannot be determined, not even in principle, from a knowledge of the periodic charge distribution of the polarized crystalline material. This establishes a fundamental difference between finite systems (e.g., molecules, clusters, etc.) and infinite periodic ones. For the first case, the dipole moment can be easily expressed in terms of the charge distribution. While for periodic systems one focuses on differences in polarization between two states of the crystal that can be connected by an adiabatic switching process [6]. The polarization difference is then equal to the integrated transient macroscopic current that flows through the sample during the switching process.

Therefore, the macroscopic polarization of an extended system is, according to the modern viewpoint, a dynamical property of the current in the adiabatic limit. The charge density is a property connected with the square modulus of the wavefunction, while the current also has a dependence on the phase. Indeed, it turns out that in the modern theory of polarization [7–9], the polarization difference is related to the Berry phase [10, 11], defined over the manifold of Bloch orbitals. The theory not only defines what polarization really is, but also proposes a powerful algorithm for computing macroscopic polarizations from first principles.

The modern theory can be equivalently reformulated using localized Wannier functions [12–15] instead of extended Bloch orbitals. The electronic contribution to the macroscopic polarization **P** is then expressed in terms of the dipole of the Wannier charge distribution associated with one unit cell. In this way, **P** is reformulated as a property of a localized charge distribution which goes back to the classic definition of polarization. However, one has to bear in mind that the phases of the Bloch orbitals are essential for building the Wannier functions. They are needed to specify how the periodic charge distribution should be decomposed into localized ones. The knowledge of the periodic charge distribution of the polarized dielectric is not enough to determine the wannier functions.

The main concepts

Here I will review the central topics uncovered in the early 1990s, often known as the modern theory of polarization. Understanding this background is necessary in order follow this chapter. The general idea is to consider the change in polarization of a crystal as it undergoes a slow change and relate this to the current that flows during this adiabatic evolution of the system. These ideas will lead to an expression for the polarization that does not take the form of an expectation value of an operator, but takes the form of a Berry phase, which is a geometrical phase property of a closed manifold (the Brillouin zone) on which a set of vectors, the occupied Bloch states, are defined.

As a classical example of a Berry phase, lett's think of the paralel transport of a vector along a loop on a shpere (for instance a compass needle carried in a car traveling on the surface of the Earth). After completing a closed path, or loop, the vector will be back at the original starting point but it will be rotated with respect to the direction it was pointing at when it started the trip. The reason for this rotation is purely geometrical topological and intrinsicaly connected to the curvature of the sphere and would not exist if the vector would be parallel transported along a flat manyfold, like a plane, or cylinder. This rotation angle is related to the integral of the curvature on the surface bounded by the loop.

Now lets see the quantum counterpart and its connection to the definition of Polarization in a solid. Let's assume that the crystal Hamiltonian H_λ depends smoothly on parameter λ and has Bloch eigenvectors obeying $H_\lambda|\psi_{\lambda,n\mathbf{k}}\rangle = E_{\lambda,n\mathbf{k}}|\psi_{\lambda,n\mathbf{k}}\rangle$ and that λ changes slowly with time, so that it is correct to use the adiabatic approximation.

Since the spatially averaged current density is just $\mathbf{j} = \frac{d\mathbf{P}}{dt} = \left(\frac{d\mathbf{P}}{d\lambda}\right)\left(\frac{d\lambda}{dt}\right)$ we can write the change in polarization during some time interval as

$$\triangle \mathbf{P} = \int \mathbf{j}(t)dt \qquad (1)$$

and is phrased in terms of the current density that is physically flowing through the crystal as the systems traverses some adiabatic path. It can also be written in terms of the parameter λ as:

$$\triangle \mathbf{P} = \int \frac{d\mathbf{P}}{d\lambda} d\lambda. \tag{2}$$

Then, from Ref.[7]

$$\frac{d\mathbf{P}}{d\lambda} = \frac{ie}{(2\pi)^3} \sum_n \int_{BZ} d\mathbf{k} \langle \nabla_\mathbf{k} u_{n\mathbf{k}} | \frac{du_{n\mathbf{k}}}{d\lambda} \rangle + c.c. \tag{3}$$

and the rate of change of polarization with λ is a property of the occupied bands only. This expression can be integrated with respect to λ to obtain

$$\mathbf{P}(\lambda) = \frac{ie}{(2\pi)^3} \sum_n \int_{BZ} d\mathbf{k} \langle u_{\lambda,n\mathbf{k}} | \nabla_\mathbf{k} | u_{\lambda,n\mathbf{k}} \rangle. \tag{4}$$

It can be verified by taking the λ derivative of both sides of Eq.4 and comparing with Eq.3. The result is independent of the particular path of $\lambda(t)$ in time, and depends only on the final value of λ as long the change is adiabatically slow. We can associate the physical polarization of state λ with $\mathbf{P}(\lambda)$ and drop the λ label. Eq. 4 can be written then as

$$\mathbf{P} = \frac{e}{(2\pi)^3} Im \sum_n \int_{BZ} d\mathbf{k} \langle u_{n\mathbf{k}} | \nabla_\mathbf{k} | u_{n\mathbf{k}} \rangle, \tag{5}$$

and this is the electronic contribution to the polarization. To obtain the total polarization it must be added the nuclear (or ionic) contribution

$$\mathbf{P}_{ion} = \frac{1}{\Omega} \sum_s Z_s \mathbf{r}_s,$$

where the sum is over atoms s having core charge Z_s and spatial position $\mathbf{r}_s$ in the unit cell of volume Ω.

The equation 5 is the main result of the modern theory of polarization It states that the electronic contribution to the polarization of crystalline insulator may be expressed as a Brillouin zone integral of an operator $i\nabla_\mathbf{k}$. However, it is not a common quantum mechanical operator, the result of its action on the wavefunctions $(i\nabla_\mathbf{k} | u_{n\mathbf{k}} \rangle)$ depends on the relative phases of the Bloch functions at different $\mathbf{k}$.

To understand the nature of the integrand in Eq.5 I want to recall the fundamental paper by Zak [16] in which he postulates the existence of a Berry phase associated with each one of the bands of a one dimensional crystalline solid. He shows how the variation of k over the entire Brillouin zone produces the appearance of a Berry phase. Moreover, he associates this phase, that is a characteristic feature of the whole band, with the band center operator [17].

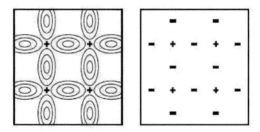

Figure 2. From Ref.[18] Mapping of the distributed charge density onto the centers of charge of the Wannier functions.

Its eigenvalues turn out to give the average positions of an electron in different bands which coincide with the symmetry centers of the space group of the solid [17] These previous ideas set the stage for the crucial meaning of the expression in Eq.5.

In three dimensions, the Brillouin zone can be regarded as a closed 3-torus obtained by the identifying points $u_{n\mathbf{k}} = u_{n,\mathbf{k}+\mathbf{G}_j}$ where $\mathbf{G}_j$ are the three primitive reciprocal lattice vectors. Then, the electronic polarization contribution of each band can be written as

$$\mathbf{P}_n = \frac{-e}{2\pi\Omega} \sum_j \Phi_{nj} \mathbf{R}_j \tag{6}$$

where $\mathbf{R}_j$ is the real space primitive translational corresponding to $\mathbf{G}_j$, and the Berry phase for band n in direction j is

$$\Phi_{nj} = -\frac{\Omega}{(2\pi)^3} Im \int_{BZ} d^3k \langle u_{n\mathbf{k}} | \mathbf{G}_j . \nabla_{\mathbf{k}} | u_{n\mathbf{k}} \rangle. \tag{7}$$

More details and discussion, including the equivalent of Eq.7 for the case of connected multiple bands, may be found in Refs.[7] and [8].

Reformulation in terms of Wannier Functions.

We can rewrite Eq.5 in terms of the Wannier functions (WF's), which brings an alternative and more intuitive way of thinking about it. The WF's and Bloch functions can be regarded as two different orthonormal representations of the same occupied Hilbert space. The WF's are localized functions $w_{n\mathbf{R}}(\mathbf{r})$ that are labeled by band n and unit cell $\mathbf{R}$ and constructed by carrying out a unitary transformation of the Bloch states $\psi_{n\mathbf{k}}$. They are constructed via a Fourier transform of the form

$$|w_{n\mathbf{R}}\rangle = \frac{\Omega}{(2\pi)^3} \int_{BZ} d\mathbf{k} e^{i\mathbf{k}.\mathbf{R}} |\psi_{n\mathbf{k}}\rangle \tag{8}$$

where the Bloch states are normalized in one unit cell. There is some freedom in the choice of the these WF's, as the transformation is not unique. In particular, a set of Bloch functions

$$|\psi_{n\mathbf{k}'}\rangle = e^{-i\beta_n(\mathbf{k})}|\psi_{n\mathbf{k}}\rangle \tag{9}$$

results in WF's $w_{n\mathbf{R}'}$ which are different from the $w_{n\mathbf{R}}$. Usually, this "gauge" is set by some criterion that keeps the WF's well localized in real space, such as the minimum quadratic spread criterion introduced by Marzari and Vanderbilt [19]. However, it is expected that any physical quantity, such as the electronic polarization arising from band n, should be invariant with respect to the phase change $\beta_n(\mathbf{k})$.

Once we have obtained the WF's, we can locate the "wannier centers" $\mathbf{r}_{n\mathbf{R}} = \langle w_{n\mathbf{R}}|\mathbf{r}|w_{n\mathbf{R}}\rangle$. It can be shown that

$$\mathbf{r}_{n\mathbf{R}} = \mathbf{R} + \sum_j \frac{\phi_{nj}}{2\pi}\mathbf{R}_j, \tag{10}$$

where ϕ_{nj} is given by Eq.7. In simple words, the location of the n'th Wannier center in the unit cell is just given by the three Berry phases ϕ_{nj} of band n in the primitive lattice vector directions $\mathbf{R}_j$. The key result, is that the polarization is just related to the Wannier centers by

$$\mathbf{P} = -\frac{e}{\Omega}\sum_n \mathbf{r}_{nO}. \tag{11}$$

The Berry Phase theory can then be regarded as providing a mapping of the distributed quantum mechanical electronic charge density onto a lattice of negative point charges $-e$ (see Fig.2).

3. Layer polarization definition

One issue that has received much attention theoretically is how to quantify the concept of *local polarization*. This can be very useful in understanding the enhancement or suppression of spontaneous polarization; or even to find new spatial patterns of polarization that would remain hidden otherwise, as is the case that I will show in a following section. It can also be essential for characterizing and understanding interface contributions to such properties.

But first I will give a review of the concept introduced in 2006 by Wu.*et al.* [20] which is the one used here. As explained in the previous section, the modern theory of polarization establishes that the polarization of a crystal is expressed in the Wannier representation as the contribution of the ionic and electronic charge:

$$\mathbf{P} = \frac{1}{\Omega}\sum_s Q_s\mathbf{R}_s - \frac{2e}{\Omega}\sum_m \mathbf{r}_m, \tag{12}$$

where s and m run over ion cores (of charge Q_s located at $\mathbf{R}_s$) and Wannier centers (of charge $-2e$ located at $\mathbf{r}_m$), respectively, in the unit cell of volume Ω.

They defined the Layer polarization (LP) along z for superlattices built from II-VI ABO_3 perovskites such as $BaTiO_3$, $SrTiO_3$, and $PbTiO_3$. For these cases, they were able to decompose the system into neutral layers (that is AO and BO_2 subunits) and define a layer polarization

$$p_j = \frac{1}{S} \sum_{s \in j} Q_s \mathbf{R}_{sz} - \frac{2e}{\Omega} \sum_{m \in j} z_m, \tag{13}$$

in which the sums are restricted to entities belonging to layer j. S is the basal cell area and we are now focusing only on z components. The LP p_j thus defined has units of dipole moment per unit area. The total polarization, with units of dipole moment per volume, is exactly related to the sum of LP's via

$$P_z = \frac{1}{c} \sum_j p_j \tag{14}$$

where $c = \frac{\Omega}{S}$ is the supercell lattice constant along z.

They propose two conditions to be able to associate a physical meaning to the LP definition: (i) resolve the arbitrariness associated with the positions of the Wannier centers, and (ii) associate without ambiguity the right number of Wannier centers to each layer.

As it will be shown in the next section, we applied this concept to thin ferroelectric films sandwiched between metal contacts. These kind of systems, in which metals and oxides coexist, bring new challenges. Condition (i) was satisfied using maximally localized wannier functions plus a disentanglement procedure (Ref.[21]) and (ii) was satisfied, even close to the metal oxide interfaces.

4. Depolarization field

When dealing with ferroelectric thin films, a finite polarization normal to the surface will give rise to a depolarizing field and a huge amount of electrostatic energy, enough to be able to suppress the ferroelectric instability. In order to preserve ferroelectricity and minimize the total energy, the depolarization field must be screened, either by free charges coming from metallic electrodes or by breaking into a domain structure. I will go into more details about this, but first a basic electrostatic background will be given.

Influence of the electrical boundary conditions

In the case of a free standing slab of a ferroelectric material, a uniform polarization P with an out of plane direction will appear and originate a surface polarization charge density

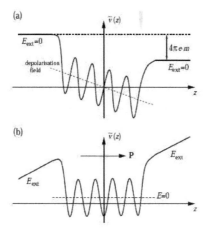

Figure 3. Schematic representation of the planar averaged electrostatic potential (fill line) of a slab with polarization perpendicular to the surface under (a)$D = 0$ boundary conditions (vanishing external field) and b) a vanishing internal electric field $E = 0$. Plannar averages are taken on the (x, y) planes parallel to the surface. Dashed lines represent an average over unit cell of the planar averages. An estimate of the macroscopic internal field inside the slab can be obtained from their slope. m stands for the dipole moment parallel to the surface normal and e for the electron charge. From Ref.[23]

$$\sigma_{pol} = \mathbf{P}.\hat{n}, \tag{15}$$

where $\hat{n}$ is a unit vector normal to the surface pointing outward. The charge density is proportional to the magnitude of the normal component of the polarization inside the material [22] and is positive one one side of the slab and negative on the other one. In order the determine the electric fields generated by the surface charge density, we need to know the electrical boundary conditions of the problem and solve the Poisson equation. For the general case, let's suppose that an external electric field E_{ext}, perpendicular to the surface, is applied in the vacuum region. There, the electric displacement $\mathbf{D} = \mathbf{E} + 4\pi\mathbf{P}$ equals the applied electrical field $D = E_{ext}$ as the polarization is zero. The normal component of D is conserved across the ferroelectric/vacuum interface,

$$E + 4\pi P = E_{ext} \tag{16}$$

Now we can see two extreme cases. One of them is that of zero external electric field, $E_{ext} = 0$, in which case the displacement vector is null while the internal field inside the slab is $E = -4\pi P$ (see Fig.3a). The absence of an external electric field and the presence of the surface polarization charges produce an internal field that points in the direction opposite to that of the polarization. Therefore, it tends to restore the paraelectric configuration and that is the reason of its name, *Depolarization field*. The coupling between the polarization and this internal field is so big, that any atomic relaxation under these circumstances will end up with the atoms back in the centrosymmetric non polar configuration.

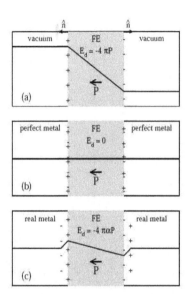

Figure 4. Depolarization field for an isolated free standing slab in the absence of an external electric field (a), and for a ferroelectric capacitor with perfect screening (ideal metal) (b) and a real metallic electrode (c) under short circuit boundary conditions. From Ref.[1]

Another case, a vanishing internal electric field $E = 0$ corresponds to a case where no field opposes the polarization and a spontaneous polarization might arise (Fig.3b). In this case, it is the continuity of the normal component electric displacement vector that establishes the value of it,

$$E_{ext} = 4\pi P_s. \tag{17}$$

These two cases illustrate how a polarization perpendicular to the surfaces can exist or not depending on the electrical boundary conditions. For simulating a real system, the mechanical boundary conditions should be included via atomic relaxations. These cases have been extensively analyzed [23–26] and have opened the ground for more realistic systems..

Metal contacts

The next level of approximation to a real case,is the case of having the ferroelectric thin film sandwiched between metal contacts (short circuit boundary conditions). In the case of a *perfect* metal, the surface charges σ_{pol} will be exactly compensated at the interfaces by the compensation charges σ_{com} and we will have a null depolarization field (see Fig.4b). In this case, both $\sigma's$ will lie in a sheet right at the interface between the metal and the ferroelectric electrode. However, in the case of *real electrodes*, the compensation of the polarization charges will be incomplete. In this case,the compensation charge σ_{com} will be spread over a finite distance inside the electrode and will create an interface dipole density at each of the ferroelectric/metal interfaces (see Fig.4c). Due to the short circuit boundary conditions, an amount of charge must flow from one interface to the other, creating a residual depolarization field inside the thin film. There are various models for expressing this field [24, 27–29], shown below is one described in Ref.[30]:

$$E_{dep} = -\frac{8\pi P \lambda}{d} \tag{18}$$

where d is the film length, and λ is the effective screening length, proportional to the spatial spread of the interface dipole.

Electrostatic energy of the thin film

In the presence of a residual depolarizing field the energy of the thin film U can be approximated by

$$E(P_0) = U(P_0) + E_{elec}(P_0), \tag{19}$$

where U is the internal energy under zero field. If we assume that the interface effects are hidden in the screening length parameter, then they only appear through the depolarization field. This is a rough approximation but we will introduce later a model that will relax this condition, but let's keep it for now. Following this, then

$$U(P_0) = mU_{FE}(P_0) \tag{20}$$

where m stands for the number of unit cells of the film. U_{FE} can be calculated from bulk DFT calculations. Its dependence on P_0 has typically the shape of a double well.

The electrostatic energy $E_{elec}(P_0)$ can be approximated by [1, 31]

$$E_{elec}(P_0) = m\Omega\left(-E_{dep}.P_0\right) + O\left(E_{dep}\right)^2 \tag{21}$$

The depolarization field can be calculated from first principles by taking the electrostatic average [32, 33] or can be expressed using Eq.18. In this case, Eq.19 can be rewritten as

$$E(P_0) = U(P_0) + \frac{8\pi\lambda P_0^2}{d} + O\left(P_0^4\right). \tag{22}$$

The electrostatic energy, the second term, is positive, meaning that the effect of the depolarization field is to suppress the ferroelectric instability by rescaling the quadratic term of the double well. This simple model has been implemented [34, 35] in order to extend the

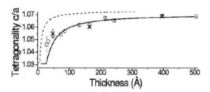

Thickness (Å)

Figure 5. Evolution of the c/a ratio with the film thickness for monodomain $PbTiO_3$ films grown epitaxially on top of Nb-doped $SrTiO_3$. With circles and squares, the experimental results. The dashed line is the phenomenological theory prediction. Solid lines correspond to the first principles model Hamiltonian results. From Ref. [35].

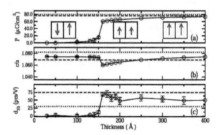

Thickness (Å)

Figure 6. Theoretical predictions of the thickness dependence of the normal average polarization P, the tetragonality c/a, and the out of plane piezoelectric constant d_{33} at room temperature for $PbTiO_3$ thin films grown on a $SrRuO_3/SrTiO_3$ substrate. Values of the quantities at the bulk level are represented by dashed lines for the unstrained configuration, and with dotted lines for a geometry under the strain imposed by the substrate. The evolution of the system, from a monodomain configuration at large thickness, (where the depolarization field E_d is small), to a 180∘ stripe domain structure in order to minimize the energy associated with E_d, is represented in the inset. From [37]

first principle model Hamiltonian to thin films. It also has been used with the result of first principle calculations in Ref.[30]. In this work, we will use it as a starting point for a toy model that would mimic the result of our first principle calculations.

4.1. Effects of the depolarization field

There are two ways to minimize the energy due to the depolarization field , i) a reduction of the polarization, ii) a reduction of the depolarization field itself. This can be seen by inspecting the first term in Eq.21.

As an example of case i) it is shown in Fig.5 the experimental results for the tetragonality in function of the film thickness for monodomain $PbTiO_3$ epitaxially grown on Nb-doped $SrTiO_3$ [35]. The polarization is strongly coupled to strain in perovskites oxides [36], therefore a reduction of the polarization must be accompanied by a reduction of the tetragonality of the system. The incomplete screening of the depolarization field is the driving force for the reduction of the polarization. The structure remains in a monodomain at all thickness.

Case ii) might arise when there are no compensation charges provided by the electrodes or when the compensation charges do not provide an efficient enough screening of the polarization charges. In this case the system might break up into 180∘ stripe domains to reduce the magnitude of the surface dipole density (see Fig.6).

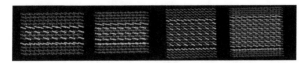

Figure 7. We simulated ferroelectric films at different thicknesses, sandwiched between metal contacts.

The reason why some systems remain in a monodomain configuration while other similar heterostructure break up into domains remains an open question that requires further clarification.

In a seminal theoretical paper, Junquera *et al.*[30] has suggested that the appearance of ferroelectricity in thin films of $BaTiO_3$, with $SrRuO_3$ contacts, should be limited to a thicknesses of more than six unit cells. Below that limit, ferroelectricity would effectively be canceled by the strength of the depolarization field (DF) inside the oxide produced by unscreened charges at the interfaces, which, in that work, were calculated for the first time in a first principles framework. More recently this picture has been broadened, and a variety of studies have shown both experimentally and with more realistic ab initio simulations that thin oxide films can indeed maintain some ferroelectricity at thicknesses even smaller than six unit cells.[37–40]

5. The hidden nature of ferroelectricity at the nanoscale

The size limit of ferroelectricity in ultrathin films can be understood using a simple electrostatics argument: as the thickness of the oxide film is reduced, the intensity of the DF is made stronger,[30] thus increasing the electrostatic energy of the system. In a ferroelectric crystal, this energy contribution can be minimized in two ways: either by breaking the polarization pattern in 180o stripe domains [41–44] to reduce the magnitude of the surface dipole density, or by a reduction of the ionic polarization while the system remains in a monodomain state.[35, 45] The final polarization pattern depends on the individual material and/or structure, and understanding which system modification will occur under what conditions is still a matter of debate.

At small dimensions, the spatial confinement and the interface details become extremely relevant since they determine the spatial localization of the electrons and thus influence directly the ferroelectric characteristics and, more in general, all the electronic properties of the system.[46] Therefore, to clarify the behavior of ferroelectric thin films in the nanometer regime it is crucial to obtain a precise description not only of the geometry and electronic properties of the film but first and foremost of the polarization profile inside the oxide.

In order to obtain a complete description of ferroelectricity at the nanoscale we have exploited the notion of layer polarization (LP). This idea, based on modern theory of polarization and the concept of maximally localized Wannier functions (see Sec.2), was recently introduced to describe the layer-by-layer modulation of polarization in ferroelectric superlattices (Sec.3) .

We have applied this method to a model metal-ferroelectric system ($BaTiO_3$/Pt) (see Fig.7) and obtained a detailed spatial profile of the polarization in the oxide region in the direction of growth that accounts not only for the ionic displacements but also for the spatial rearrangement of the electron density. It is important to note that using more standard methods to evaluate macroscopic polarization, this information would have remained hidden

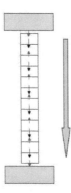

Figure 8. The thin film oxide develops a ferrielectric [47] pattern of polarization that exhibits antiferroelectric properties along the growth axis and with a net total polarization. This pattern would remain hidden by using more standard methods for calculating the local polarization.

by the averaging procedure, or would have produced misleading results. In fact, by considering exclusively the ionic displacements (rumpling) as a measure of polarization, we do not account for the full electronic contributions to the local polarization and, depending on the system size, we might obtain results that would not describe accurately the physical behavior of the thin film.

Here, we show that the ferroelectric structures associated with small length-scales are more complex than previously thought, mainly due to the redistribution of the electrons under the constraints imposed by the interfaces. In particular, the oxide develops a ferrielectric [47] pattern of polarization that exhibit antiferroelectric properties along the growth axis and with a net total polarization (see Fig.8).

This unbalanced dipole structure is particularly evident in Fig.10A, where we show how positive rumpling on all the atomic layers of a thin $BaTiO_3$ film correspond actually to layer dipoles of opposite signs. This is a consequence of the interplay between the orientation and magnitude of the dipoles with the DF, their mutual interaction and the nature of the interfaces. A simple analytic model of a ferroelectric thin film, where these effects are explicitly taken into account, can capture all these features and it will be discussed in detail at the end.

We have simulated thin films of $BaTiO_3$ between Pt metal contacts in a (001) stack, where the oxide is terminated with a BaO plane at both interfaces. In this configuration, the metal atoms are directly bonded to the O atoms of the oxide plane.[30, 48] The different supercell constructed in this way can be labeled as Pt/(BaO-TiO2)m-BaO/Pt with m=1,2,4,6. Nine atomic planes of Pt have been found to be sufficient to simulate the contacts under short-circuit boundary conditions. The in plane lattice constant of the supercell is set equal to that of $BaTiO_3$ at 0 K (3.991)[38], and kept it fixed in all calculations while all other geometrical parameters (atomic positions and intra-layer distances) were fully relaxed. All simulations have been performed using Density Functional Theory (DFT) within the Local Density Approximation, with ultrasoft pseudopotentials and a plane waves basis set.

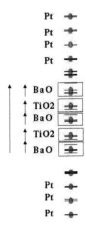

Figure 9. The set of MLWFs so obtained cluster around each atomic layer in sets containing the correct number of Wannier function needed to keep the layer charge neutral. Here we can see the z position (horizontal lines lines) of the centers of the Wannier functions and the position of the ions (indicated with circles). To calculate the LP of each layer, we use Eq.12.

We used the expression of Eq.12 to calculate the LP for each plane of the thin film. A is the in plane area of the cell, and we are only interested on the z components (parallel to the growth direction). It is important to stress that in the definition of the LP we have explicitly included the full ionic and electronic information necessary to define unambiguously the local value of the polarization. The essential step for the evaluation of the layer polarizations p_j is the determination of the Wannier centers z_m. We used the maximally localized Wannier functions (MLWFs) algorithm originally proposed by Marzari and Vanderbilt[19] as implemented in the WanT code[1]. Since the valence bands of the oxide and the bands of the metal are mixed in the full supercell calculation, a disentanglement procedure (Ref.[21]) was applied before starting the localization algorithm[19, 49].

The set of MLWFs so obtained cluster around each atomic layer in sets containing the correct number of Wannier function needed to keep the layer charge neutral (Fig.9). This is a crucial condition for the validity of the definition of LP's in Eq.12 and might not be attainable in systems with more covalent nature.[20] In fact, the meaningful use of this concept is based on the condition of being able to decompose the system into neutral layers (TiO and BaO2 units in the case of perovskite BaTiO3) and resolve the arbitrariness associated with the positions of the Wannier centers. This allows us to associate each sets of Wannier centers to the corresponding crystal plane without ambiguity. By doing so we go back to the classic idea of defining dipoles in spatial neutral units, along the lines of the Clausius-Mossotti limit.v Moreover, we can associate each LP value p_j with its correspondent (dimensionally correct) dipole $p_j A$, which gives precise information about the orientation and magnitude of the atomic layer dipoles inside the crystal.

---

[1] code by A. Ferretti, B. Bonferroni, A. Calzolari, and M. Buongiorno Nardelli, (http://www.wannier-transport.org). In the non-self-consistent calculation for obtaining the MLWFs we used MP k meshes 2x2xN with N=16,14,12,10 for m=1,2,4,6 respectively. I. Souza, N. Marzari, and D. Vanderbilt, Phys. Rev. B 65, 035109 (2001). ote

The LP profile calculated for the metal-oxide systems provides a detailed local description of the polarization, which is superior and even contradictory to the information obtained from empirical evaluations based solely on ionic displacements. As an illustration, in Fig.10A we display the rumpling profile (cation-oxygen normal distance within each atomic layer.) for a thin film comprised of six BaTiO$_3$ unit cells (m6). From the ionic displacement profile one would be tempted to conclude that the film is in a ferroelectric domain configuration, since the rumpling values (and hence the *classical* local dipoles) associated to each plane are positive. Instead, if the correspondent LP profile is analyzed, a completely different picture emerges, where the main feature is an alternation of positive and negative values of the LP in the individual atomic layers of the oxide film. This is clearly shown in Fig.10B, where we plot the LPs values computed for m6. Here the central BaO planes (the BaO planes at the interface behave differently and will be discussed later) have positive dipoles while the TiO2 planes have negative ones. The first are aligned with the direction of the DF, while latter oppose it. A schematic illustration of the associated dipoles is drawn above the plot. This result implies that what could have been interpreted as a ferroelectric domain using standard geometrical information, in fact displays a much more complex spatial pattern of polarization. We believe that this effect is comparable to the onset of the formation of two-dimensional domain islands commonly observed both theoretically and experimentally in ferroelectric films.[37] In that case, the domains are composed of adjacent 180o domains in order to minimize the electrostatic energy associated with the DF. Here, our results show that the system can reduce its electrostatic energy in an alternative way: the energetic interplay between the dipoles, the DF and the interface bonding drives the individual atomic layer dipoles to arrange themselves in an uncompensated dipole pattern along the growth direction (see inset of Fig.10A), or, in other words, the system undergoes a ferroelectric-to-ferrielectric phase transition. Similarly to the formation of two-dimensional islands, we also predict the existence of a critical thickness (CT) above which the system exists in a true ferroelectric domain with all the dipoles pointing in the same direction (see Fig8B, where the LP's values for bulk BaTiO$_3$ are shown with triangles).

As the thickness is reduced, the magnitude of all the dipoles is diminished by the increase of the DF that opposes them. Due to the different physical properties of consecutive layers (BaO and TiO2), their associated dipoles reduce their magnitude at different rates. At the CT, the BaO layers will have null dipoles and eventually they will flip layer polarization direction, aligning with the DF in order to minimize the associated internal energy and creating a ferrielectric dipole pattern (FDP).

The formation and characteristics of the FDP are determined by the bonding properties of the interfaces that pin the LP values of the outer layers of the oxide film. At small film thicknesses the interface effects become dominant and induce the overall spatial variation of the LP's that determines the orientation and magnitude of the layer dipoles along the structure. We can see how the FDP observed in the m6 geometry is increasingly influenced by the locality of the interfaces as the thickness is reduced to m4, m2 and m1 (see Fig.10C,D,E respectively). The middle crystal planes become more and more influenced by the interface environment, while the FDP becomes irregular, and finally disappear when the film is comprised by just 3 layers (i.e. one unit cell, m1), leaving instead a centro-symmetric paraelectric structure (Fig.10E). It is important to note that exclusively electronic effects drive this series of regime transitions. In fact, the ionic polarization, directly proportional to the ionic rumpling, remains almost

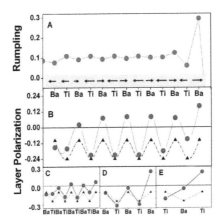

Figure 10. A. Rumpling (cation-oxygen perpendicular distance (in Å) within each atomic layer) profile for six unit cells (m6) of BaTiO₃ sandwiched between Pt contacts. (on the horizontal axis Ti and Ba indicate the position of the individual TiO2 and BaO planes). B. Solid circles (blue on line): layer polarization (LP) profile for the same system m6 in units of $10^{-}10C/m$. The values for the equivalent planes in bulk BaTiO₃ are shown with black triangles. The orientation of the individual layer dipoles is displayed by arrows in the bottom of the panel. C.D.E. LP profiles for m4, m2 and m1. As the thickness is reduced the FDP is modified by the increasingly dominant interface effects.

constant for all the thicknesses considered, and it is only the electronic polarization that changes its values in the different systems.

These results show that the ferroelectric response of a thin film is indeed critically influenced by the interface properties of the system and that the analysis of the local variation of layer polarization captures completely the physical characteristics of the ferroelectric. As the size of the film is increased, the DF decreases and the dipoles that are oriented along the DF direction get smaller, while the interface effects lose relevance. At the critical size, the dipoles flip direction and a ferroelectric domain structure is established again. Eventually, the dipoles will relax to the bulk structure, showed with triangles in Fig.10

Model

We further illustrate these physical principles by developing a simple classical model of a ferroelectric thin film that reproduces the appearance of a FDP. We extended the model introduced in Ref. ([30]) by discretizing the thin film along the growth direction in each of its component unit cells in order to introduce a local spatial description of the polarization with the introduction of the *local polarization*, P_i. Each individual crystal unit cell is comprised of two consecutive layers of BaO and TiO2, therefore P_i could be obtained by adding each individual LP and dividing by the lattice constant in the stacking direction, c. The full polarization profile of the film is completely specified by the full set of LPs $(p_1, p_2, ..., p_{2M-1}, p_{2M})$ associated with each crystal plane of the oxide film (see Fig.11). We can write the total energy of the thin film composed by M unit cells as:

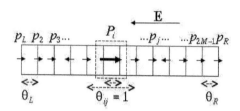

Figure 11. Each individual crystal unit cell is comprised of two consecutive layers of BaO and TiO2. A local polarization P_i can be associated with it. It is calculated by adding both individual LP's and dividing by the lattice constant in the stacking direction, c. We can associate a dipole $p_i A$ to each plane i, separated from its next neighbors by distance d_{ij}. The full polarization profile of the film is completely specified by the full set of dipoles $(p_L, p_2, ..., p_{2M-1}, p_R)$. The LP at the interfaces (p_L and p_R), are kept fixed. Two distinct interaction parameters θ_L and θ_R, establish the locality of the interface environment.[50] The interaction parameter θ_{ij} for the internal dipoles is arbitrarily chosen to be unity.

$$E = \sum_{i=1}^{M}(-aP_i^2 + bP_i^4) + \sum_{i=1}^{M}(-\Omega P_i.E) + U_{int}(p_L, p_2, ..., p_{2M-1}, p_R), \qquad (23)$$

where each term of the first sum accounts for the internal energy of the individual unit cells under zero electric field. The parameters a and b are obtained from an ab initio calculation of the total energy of a BaTiO$_3$ bulk for different ionic displacements along the soft mode.[51] The contribution from the depolarization field E is included in the second sum [22, 30] where Ω is the volume of the unit cell. Each one of its terms favors energetically the local dipoles that follow the same direction than the DF. The third sum represents the classical electrostatic energy for a series of dipoles separated by a distance d_{ij},

$$U_{int} = -\frac{A^2}{8\pi\epsilon} \sum_{i,j=1}^{2M} \frac{\theta_{ij} p_j p_i}{d_{ij}^3} \qquad (24)$$

The above sum can be separated in three contributions: two that account for the interaction between the interface dipoles with the internal ones, and another one that accounts for the mutual electrostatic interaction between the internal dipoles. We have chosen to assign a fixed magnitude and direction to the interface layer dipoles (p_L and p_R), with two distinct interaction parameters θ_L and θ_R, thus establishing the locality of the interface environment.[50] The interaction parameter θ_{ij} for the internal dipoles is arbitrarily chosen to be unity.

With the constraint imposed by the simplicity of the model, we look for the energetically more favorable spatial polarization profile as a function of the film thickness. The spatial profile of polarization that characterize the film can be expressed as the vector $(p_L, p_{TiO_2}, p_{BaO}, p_{TiO_2}, ..., p_{BaO}, p_{TiO_2}, p_R)$ where a ferroelectric domain and a FDP will have all dipoles parallel or antiparallel respectively. Note that the total energy per unit cell ($E' = \frac{E}{M}$) an be expressed as function of only two single variables p_{TiO_2} and p_{BaO} in the form:

$$E'(p_{TiO_2}, p_{BaO}) = -a \left(\frac{p_{TiO_2} + p_{BaO}}{c} \right)^2 + b \left(\frac{p_{TiO_2} + p_{BaO}}{c} \right)^4 + \Omega E \left(\frac{p_{TiO_2} + p_{BaO}}{c} \right)$$

$$-\frac{A^2}{4\pi\epsilon} \left(\frac{\alpha p_{BaO}}{Mc^3} + \frac{p_R p_L}{M(2M-1)c^3} + \frac{1}{M} \sum_{i=3}^{2M-1} \frac{p_L p_i}{d_{Lj}^3} + \frac{1}{M} \sum_{i=3}^{2M-1} \frac{p_i p_R}{d_{iR}^3} + \frac{1}{2M} \sum_{i,j=4}^{2M-2} \frac{p_i p_j}{d_{ij}^3} \right) \tag{25}$$

where we used Eqs.2324 and we define an overall interface parameter $\alpha = (\theta_R p_R + \theta_L p_L)$. Note that the thickness of the film is given by the number of unit cells M and directly affects the magnitude of the depolarization field E. The DF is calculated from an ab initio calculation for a supercell with the same size M and using a well-established averaging technique for the electrostatic potential inside the film.[50]

As $M \to \infty$, the terms in Eq.25 that describe the mutual interaction between the dipoles converge to constant values, the DF vanishes and only the first two terms remain in E?, recovering the bulk properties. Only for small M these terms contribute appreciably to the total energy of the system. This is true in particular for the interface term that contains the parameter (fourth term in Eq. 25) that carries the information that defines the interfaces. We can estimate the interface parameters θ_R, θ_L, p_L, p_R from our ab initio simulations[2] and we assume that their numerical values will not depend upon the film thickness (as observed in the actual calculations, at least to first order).

As we vary the thickness of the ferroelectric film, we observe a change of the total energy landscape consequence of the interplay between the different energetic contributions. This is clearly seen in Fig. 12, where we show contour plots of $E'(p_{TiO_2}, p_{BaO})$ for different values of M. On the left, a diagram with the correspondent spatial distribution of dipoles is indicated. For thick films, the spatial distribution of dipoles that minimizes the total energy forms a domain (both LP's in the BaO and TiO2 planes have the same sign), as can be seen in the position of the minimum in the contour plot in Fig.12A. As the number of unit cells is lowered, the minimum shifts to lower values of p_{BaO} until at a critical thickness (Fig. 12B) it becomes zero. Further reduction of the number of layers M flips only the dipoles belonging to BaO planes thus establishing a FDP (see Fig. 12D). In conclusion, this simple toy model captures the general physics of the thin film, and illustrates the intimate relation between the interface characteristics, the thickness of the film, and the existence of an FDP.

Thus, by exploiting the concept of layer polarization in the description of ferroelectric thin films between metal contacts, we have been able to obtain detailed information on the modulation of polarization at the nanoscale and to understand the constraining effects of the interfaces in the determination of the ferroelectric response of the system. Our results shows that when film thicknesses reach a critical value, the ferroelectric system responds via a transition from the bulk ferroelectric structure to a ferrielectric antidipole pattern, where individual atomic layers acquire uncompensated opposing dipoles. This state arises as

---

[2] The DF does not change much for the films thicknesses considered in this work. Therefore the value used as input for the model was calculated from the m6 averaged electrostatic potential ($E = 2.1410^14V/m$). The values for both interface layer dipoles used are ($(p_R + p_L)/c = 1.0C/m^2$). In the model we also chose $c/a = 0.98$, respecting the tetragonal structure of the internal cells of the oxide film.

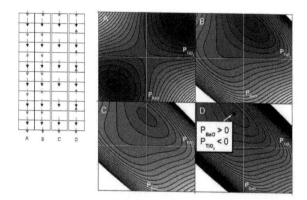

Figure 12. Energy landscape $E'(p_{TiO_2}, p_{BaO})$ for different values of θ/M (film thickness). The interaction parameters for both interfaces were considered equal, $theta_R = \theta_L = \theta$ and were kept fixed as the film size M is varied. The remaining parameters of the model have been derived from our ab initio calculations. A. $\theta/M = 0$, bulk limit where the usual double well potential produces two equivalent minima with all the layer dipoles parallel. B. $\theta/M = 3$, as the thickness is reduced the interface effects gain relevance and start to modify the energy landscape. The layer dipoles associated with the BaO planes reduce their magnitude in order to minimize the total energy. C. At $\theta/M = 5$ the system reaches a critical thickness where these dipoles become zero and eventually flip orientation. D. For $\theta/M = 10$ the system display a ferrielectric dipole pattern.

consequence of the complex energetic competition between the interface effects, the DF, and orientation and mutual interaction of the layer dipoles. The appearance of this particular ferrielectric state can be understood using a simple phenomenological model where the interface effects are explicitly taken into account.

These results suggest the possibility that such FDPs could be the normal state of a ferroelectric thin film at the nanoscale, even combined with the formation of two-dimensional island domains. It would be tempting to link the formation of such FDPs to the appearance or not of two-dimensional island below the critical thickness, and to understand the ultimate dependence upon the detailed interface structure and the nature of the metal contact. The next section will explore in detail the former point.

6. Tuning of polarization in metal-ferroelectric junctions

In this section we study the paradigmatic case of a BaTiO$_3$ film between Pt contacts, already introduced in the previous section, where we modify the ferroelectric properties by a selective control of the chemical species present at the interface. In particular, by inserting a single layer of a different metal (Au,Cu) we demonstrate that we are able to tune the polarization of the ferroelectric film via the modification of the screening properties of the composite metal contacts and through the change in the geometrical constraints at the interface.

6.1. Methods and discussion

We have simulated thin films of BaTiO$_3$ between Pt metal contacts in a (001) supercell. The oxide is terminated with a BaO plane at both interfaces, with the metal atoms directly bonded to the O atoms of the oxide planei,ii. The supercells constructed in this way can be labeled as $Pt/(M)/(BaO - TiO2)_m - BaO/(M)/Pt$ with m=1,2,4,6 and M=(Pt, Au or Cu). Nine

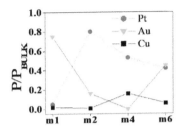

Figure 13. Normalized polarization for different number of oxide layers and interface phases. The local polarization is calculated for each oxide layer, added up and divided by the BaTiO$_3$ bulk polarization calculated in the same way with equivalent numbers of layers.

atomic metal planes (including the intralayer) have been found to be sufficient to simulate the contacts under short-circuit boundary conditions. We used the experimental in-plane lattice parameter of $BaTiO_3$ (3.99Å), and kept it fixed in all calculations. All simulations have been performed using Density Functional Theory (DFT) within the Local Density Approximation, using ultrasoft pseudopotentials and a plane waves basis set.[52]

A detailed analysis of the polarization at the nanoscale is critical for a complete description of the physical properties of the ferroelectric thin film. We have used Modern Theory of Polarization [6, 8, 18, 53, 54] and in particular the concept of layer polarization (LP)[20] to evaluate the polarization of the different structures and extract the information on the local profile of polarization at the nanoscale.

Turning to the results, in Fig.13 we show the total polarization (normalized to the bulk value) in BaTiO$_3$ films of different thickness between composite metal contacts. These results clearly elucidate our claim: the polarization of the film is critically affected by the contact geometry at all the thicknesses we have considered and a residual polarization can be observed in films as thin as one unit cell.

In fact, a one unit cell thick BaTiO$_3$ film between plain Pt contacts is still in a ferroelectric state with a polarization 10% of the ideal bulk, and the introduction of a Au intralayer enhances this value up to 70% of that! In contrast, thicker (6 unit cells) oxide films display similar ferroelectric characteristics with 50% of the bulk polarization, a clear indication of the rapid decay of the interface effects with the film thickness. It is worth to note that at variance with the previous cases, a Cu intralayer induces a paraelectric (or almost paraelectric) behavior for all thicknesses. We will come back to this point later in the discussion.

In order to understand better the behavior of ferroelectricity in such ultrathin films, we computed the layer-resolved spatial profile of the polarization, which is quantified by values of the layer polarization along [3] the structure. This is shown in Fig.14 for m4 films. When only Pt is present at the contact, the oxide develops a pattern of polarization composed of

---

[3] The local polarization is defined as $P_j = \frac{p_j}{c_j}$ for each layer j where c_j is half the distance between neighboring cations and p_i is the calculated layer polarization[20]. For the outermost layers, we had to make a somewhat arbitrary choice for c_j and used the distance between the outermost cation and the opposite metal plus half the distance between the cation and the one belonging to the second oxide layer.

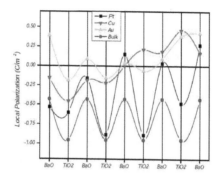

Figure 14. Local Polarization for a unit cell of size m4. Only one layer of Cu at the interface betwen the BaTiO₃ and Pt is enough to drive the system from a ferroelectric structure (squared dots) into a paraelectric one (red triangular dots).

consecutive alternated signs along the direction perpendicular to the interface (see Section 5) From the behavior of the local polarization it is clear that while the pristine Pt (curve with black squares) interface display a distinct ferroelectric behavior, a single layer of Cu (red triangular dots) at the interface between BaTiO₃ and Pt is sufficient to stabilize the system in a paraelectric (non polar) structure (the character of the structure correlates with the symmetry of the polarization profile: an asymmetric pattern correspond to a polar geometry and hence to a ferroelectric behavior; a centro-symmetric pattern on the contrary gives rise to a paraelectric behavior). If we exchange Cu for Au (curve with green triangles), the polarization is restored, although smaller that in the Pt case, and the whole polarization profile is substantially changed.

These effects vary as the size of the film changes. In thinner films the local interface details have strong effects, while they are smoothed out in thicker systems. This can be seen in Fig.15 A-D where we have plotted the LP profile for different sizes of the oxide and different metal intralayers at the interfaces. In general we notice that the Pt contact maintain the system in a ferroelectric state for all the sizes although the polarization profiles vary greatly with the system size. The same effect is observed with the addition of an Au intralayer. These changes are more noticeable in the thinner systems (m1 and m2) while for the thicker samples both systems share a similar polarization profile with a higher overall polarization in the Au case. This behavior can be directly associated with the screening of the depolarization field [4] at the metallic contacts. We have estimated the depolarization field from the macroscopic average of the electrostatic potential.[30] Indeed we observed a decrease of the DF with the introduction of the Au layer. This clearly demonstrate that Au has better screening properties compared to Pt and in consequence a Au intralayer enhances the overall polarization of the system. A similar reasoning is more difficult to do in the smaller systems due to the strong asymmetry

---

[4] We used the maximally localized Wannier functions (N. Marzari, and D. Vanderbilt, Phys. Rev. B 56, 12847 (1997)) as implemented in the WanT code (A. Ferretti, B. Bonferroni, A. Calzolari, and M. Buongiorno Nardelli, http://www.wannier-transport.org) for the determination of the Wannier centers. The disentanglement procedure (I. Souza, N. Marzari, and D. Vanderbilt, Phys. Rev. B 65, 035109 (2001)) was applied before starting the localization algorithm, since the bands of the metal and the valence bands of the oxide are mixed in the full supercell calculation.

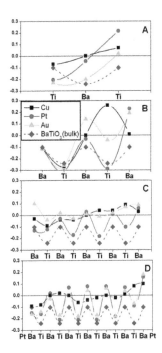

Figure 15. Layer polarization for different metal interfaces and films thicknesses A)m1, B)m2, C)m4, D)m6. As reference, the LPs for bulk BaTiO₃ are indicated with blue circles and dotted lines for each case. Metals intralayers used are Cu, Au, Pt. Case of m6 with Cu layer: this result is in fact an artificial effect due to the known LDA underestimation of the band gap. The Fermi level of the system overlaps the conduction band and starts filling states that should be empty. We expect the same issue for $m > m6$, or with the use of metals with a higher work function. The calculations for $m < m6$ are be correct under the DFT?LDA framework, as the Fermi level is safely below the conduction band.

introduced by the proximity of the interfaces and the uncertainty in the evaluation of the macroscopic averages. However, the qualitative behavior remains the same.

Contrary to Pt and Au, Cu intralayer do not stabilize any ferroelectric distortion at any thickness . This is a strong indication that Cu screening is weaker and not sufficient to reduce the depolarization field inside the oxide.

The above behavior is obviously correlated with the redistribution of charge across the interfaces in the different cases. The latter is quantified by the modification of the band alignment induced by the metal intralayer, and its influence on the screening properties of the metal contact. In fact, the knowledge of the band alignment, or Schottky barrier Height (SBH), allows us to define the properties of the interface phase between the metal and oxide. In a previous work,[51] we have demonstrated the correlation between the local structure of the interface composed by a crystalline oxide (BaO) and a d metal, and how we can tune the SBH by controlling the relative overlap of the local density of states of the different atoms close to the interface. Indeed we have found an almost complete similarity between that BaO/metal interface and the $BaTiO_3$/metal interface of this work. In both cases we considered a BaO terminated oxide slabs, that show very similar characteristics (as also indicated by the fact that the interfaces between BaO and $BaTiO_3$ have almost zero valence band offset[52]).

Indeed, we find similar behaviors for the band offsets of the interfaces BaO/M/Pd (see Ref. [52]) and $BaTiO_3$/M/Pt when Au, Cu and Pt are used as interlayer M. Furthermore, the SBH for the $BaTiO_3$/metal interfaces follow the same ascending order for each metal intralayer (Au (0.8 eV) < Pt (1.1 eV) < Cu (1.4 eV)) in both systems (values given correspond to the case of m4)[5] This trend suggests a relation between the band offsets of the interfaces and the screening properties of the metals, that is to be expected given the strong correlation between the charge transfer at the interface and the SBH.[51] In particular, if a strong dipole is established at the interface, due to a high SBH, as in the case of Cu, fewer charges will be available for screening the depolarization field and the system will not support a ferroelectric distortion. Following Ref. [51] we can state the following phenomenological rule: metal intralayers that reduce the SBH at the interfaces will enhance the screening properties of the contact and stabilize the ferroelectric distortion.

We have used Density Functinal Theory and the layer polarization concept to analyze the effect of the interface structure on thin ferroelectric films between metal contacts. As the size of the film is reduced, the interface effects become strong and influence dramatically the spatial polarization profile of the system. We monitored these changes as different metal layers were introduced at the interface, modifying the local structure of the interface, the band alignment and in turn affecting the screening nature of the metal contact. In this way we are able to tune the ferroelectric state by carefully choosing the metal interlayer.

Author details

Matías Núñez

Consejo Nacional de Investigaciones Científicas y Técnicas (CONICET), Argentina
Centro Atomico Bariloche, U-A Tecnologia de Materiales y Dispositivos, Division Materiales Nuclkheares, Argentina
Instituto de Ciencias Basicas, Universidad Nacional de Cuyo, Argentina
Previously at North Carolina State University, Raeligh, North Carolina, USA

References

[1] P. Ghosez and J. Junquera.   First-Principles Modeling of Ferroelectric Oxide Nanostructures. *ArXiv Condensed Matter e-prints*, May 2006.

[2] L. D. Landau and E. M. Lifshitz. *Electrodynamics of Continuous Media*. Pergamon Press, Oxford, 1984.

[3] C. Kittel. *Introduction to Solid State Physics*. Wiley, New York, 7 edition, 1996.

---

[5] We obtained the SBH values of the $BaTiO_3$/metal interfaces by calculating the energy difference between the Fermi level and the upper edge of the valence band, by direct inspection of the LDOS of the systems projected on of each crystal plane.

[4] N. W. Ashcroft and N. D. Mermin. *Solid State Physics*. Saunders, Philadelphia, 1976.

[5] Richard M. Martin. Comment on calculations of electric polarization in crystals. *Phys. Rev. B*, 9(4):1998–1999, Feb 1974.

[6] R. Resta. Theory of the electric polarization in crystals. *Ferroelectrics*, 136, 1992.

[7] R.D. King-Smith and David Vanderbilt. Theory of polarization of crystalline solids. *Phys. Rev. B*, 47(3):1651–1654, Jan 1993.

[8] Raffaele Resta. Macroscopic polarization in crystalline dielectrics: the geometric phase approach. *Rev. Mod. Phys.*, 66(3):899–915, Jul 1994.

[9] Raffaele Resta. Quantum-mechanical position operator in extended systems. *Phys. Rev. Lett.*, 80(9):1800–1803, Mar 1998.

[10] A. Shapere and F. Wilczek. *Geometric Phases in Physics*. World Scientific, Singapore, 1989.

[11] Raffaele Resta. Manifestations of berry's phase in molecules and condensed matter. *Journal of Physics: Condensed Matter*, 12(9):R107–R143, 2000.

[12] Gregory H. Wannier. The structure of electronic excitation levels in insulating crystals. *Phys. Rev.*, 52(3):191–197, Aug 1937.

[13] W. Kohn. Analytic properties of bloch waves and wannier functions. *Phys. Rev.*, 115(4):809–821, Aug 1959.

[14] E.I. Blount. Formalism of band theory. *Solid State Physics*, 13(305), 1962.

[15] Nicola Marzari and David Vanderbilt. Maximally localized generalized wannier functions for composite energy bands. *Phys. Rev. B*, 56(20):12847–12865, Nov 1997.

[16] J. Zak. Berry's phase for energy bands in solids. *Phys. Rev. Lett.*, 62(23):2747–2750, Jun 1989.

[17] J. Zak. Band center—a conserved quantity in solids. *Phys. Rev. Lett.*, 48(5):359–362, Feb 1982.

[18] D. Vanderbilt and R. Resta. *Conceptual foundations of materials properties: A standard model for calculation of ground- and excited-state properties.*, chapter Quantum electrostatics of insulators: Polarization, Wannier functions, and electric fields. Elsevier, The Netherlands, 2006.

[19] N. Marzari and D. Vanderbilt. Maximally localized generalized wannier functions for composite energy bands. *Phys. Rev. B*, 56(20):12847–12865, 1997.

[20] X. Wu, O. Dieguez, K. M. Rabe, and D. Vanderbilt. . *Phys. Rev. Lett.*, 97(107602), 2006.

[21] I. Souza, N. Marzari, and D. Vanderbilt. Maximally localized wannier functions for entangled energy bands. *Phys. Rev. B*, 65(035109), 2001.

[22] J.D. Jackson. *Classical electrodynamics*. Wiley, 3 edition, 1998.

[23] B. Meyer and David Vanderbilt. Ab initio study of *batio3* and *pbtio3* surfaces in external electric fields. *Phys. Rev. B*, 63(20):205426, May 2001.

[24] Ph. Ghosez and K. M. Rabe. Microscopic model of ferroelectricity in stress-free pbtio[sub 3] ultrathin films. *Applied Physics Letters*, 76(19):2767–2769, 2000.

[25] T. Maruyama, M. Saitoh, I. Sakai, T. Hidaka, Y. Yano, and T. Noguchi. Growth and characterization of 10-nm-thick c-axis oriented epitaxial pbzr[sub 0.25]ti[sub 0.75]o[sub 3] thin films on (100)si substrate. *Applied Physics Letters*, 73(24):3524–3526, 1998.

[26] Lennart Bengtsson. Dipole correction for surface supercell calculations. *Phys. Rev. B*, 59(19):12301–12304, May 1999.

[27] Batra I. P. and Silverman B. D. *Sol. State Comm.*, 11(291), 1972.

[28] P.B.Littlewood M.Dawber1, P.Chandra and J.F. Scott. Depolarization corrections to the coercive field in thin-film ferroelectric. *J. Phys.: Condens. Matter*, 15:L393–L398, 2003.

[29] A. M. Bratkovsky and A. P. Levanyuk. Very large dielectric response of thin ferroelectric films with the dead layers. *Phys. Rev. B*, 63(13):132103, Mar 2001.

[30] J. Junquera and P. Ghosez. . *Nature*, 422(506), 2003.

[31] Na Sai, Karin M. Rabe, and David Vanderbilt. Theory of structural response to macroscopic electric fields in ferroelectric systems. *Phys. Rev. B*, 66(10):104108, Sep 2002.

[32] Alfonso Baldereschi, Stefano Baroni, and Raffaele Resta. Band offsets in lattice-matched heterojunctions: A model and first-principles calculations for gaas/alas. *Phys. Rev. Lett.*, 61(6):734–737, Aug 1988.

[33] L. Colombo, R. Resta, and S. Baroni. Valence-band offsets at strained si/ge interfaces. *Phys. Rev. B*, 44(11):5572–5579, Sep 1991.

[34] Igor Kornev, Huaxiang Fu, and L. Bellaiche. Ultrathin films of ferroelectric solid solutions under a residual depolarizing field. *Phys. Rev. Lett.*, 93(19):196104, Nov 2004.

[35] Celine Lichtensteiger, Jean-Marc Triscone, Javier Junquera, and Philippe Ghosez. Ferroelectricity and tetragonality in ultrathin pbtio[sub 3] films. *Physical Review Letters*, 94(4):047603, 2005.

[36] R.E. Cohen. *Nature*, 358(136), 1992.

[37] V. Nagarajan, J. Junquera, J. Q. He, C. L. Jia, R. Waser, K. Lee, Y. K. Kim, S. Baik, T. Zhao, R. Ramesh, Ph. Ghosez, and K. M. Rabe. Scaling of structure and electrical properties in ultrathin epitaxial ferroelectric heterostructures. *Journal of Applied Physics*, 100(5):051609, 2006.

[38] C.-G. Duan, R.F. Sabirianov, W.-N. Mei, S.S. Jaswal, and E.Y. Tsymbal. Interface effect on ferroelectricity at the nanoscale. *Nano Letters*, 6(3):483–487, 2006.

[39] G. Gerra, A. K. Tagantsev, N. Setter, and K. Parlinski Parlinski. Ionic Polarizability of Conductive Metal Oxides and Critical Thickness for Ferroelectricity in BaTiO3. *Physical Review Letters*, 96(10):107603–+, March 2006.

[40] N. Sai, A. M. Kolpak, and A. M. Rappe. . *Phys. Rev.B*, 72(020101(R)), 2005.

[41] Toshito Mitsui and Jiro Furuichi. Domain structure of rochelle salt and k*h*2po4. *Phys. Rev.*, 90(2):193–202, Apr 1953.

[42] Zhongqing Wu, Ningdong Huang, Zhirong Liu, Jian Wu, Wenhui Duan, Bing-Lin Gu, and Xiao-Wen Zhang. Ferroelectricity in pb (zr0.5 ti0.5) o3 thin films: Critical thickness and 180o stripe domains. *Phys. Rev. B*, 70(10):104108, Sep 2004.

[43] M. Sepliarsky, S. R. Phillpot, D. Wolf, M. G. Stachiotti, and R. L. Migoni. Atomic-level simulation of ferroelectricity in perovskite solid solutions. *Applied Physics Letters*, 76(26):3986–3988, 2000.

[44] S. K. Streiffer, J. A. Eastman, D. D. Fong, Carol Thompson, A. Munkholm, M. V. Ramana Murty, O. Auciello, G. R. Bai, and G. B. Stephenson. Observation of nanoscale 180° stripe domains in ferroelectric *pbtio*3 thin films. *Phys. Rev. Lett.*, 89(6):067601, Jul 2002.

[45] Y. S. Kim, D. H. Kim, J. D. Kim, Y. J. Chang, T. W. Noh, J. H. Kong, K. Char, Y. D. Park, S. D. Bu, J.-G. Yoon, and J.-S. Chung. Critical thickness of ultrathin ferroelectric batio[sub 3] films. *Applied Physics Letters*, 86(10):102907, 2005.

[46] R. A. McKee, F. J. Walker, M. Buongiorno Nardelli, W. A. Shelton, and G. M. Stocks. The Interface Phase and the Schottky Barrier for a Crystalline Dielectric on Silicon. *Science*, 300(5626):1726–1730, 2003.

[47] C. F. Pulvari. *Phys. Rev.*, 120(1670), 1960.

[48] Nicola A. Spaldin. MATERIALS SCIENCE: Fundamental Size Limits in Ferroelectricity. *Science*, 304(5677):1606–1607, 2004.

[49] K.S. Thygensen, L. B. Hansen, and K. W. Jacobsen. Partly Occupied Wannier Functions. *Phys. Rev. Lett.*, 94(026405), 2005.

[50] S. Baroni, R. Resta, A. Baldereschi, and M. Peressi. *Spectroscopy of Semiconductor Microstructures*, chapter "Can we tune the band offset at semiconductor heterojunctions?", pages 251–271. Plenum, London, 1989.

[51] Matias Nunez and Marco Buongiorno Nardelli. First-principles theory of metal-alkaline earth oxide interfaces. *Physical Review B (Condensed Matter and Materials Physics)*, 73(23):235422, 2006.

[52] J. Junquera, M. Zimmerman, P. Orejon, and P. Ghosez. First-principles calculation of the band offset at BaO/BaTiO3 and SrO/SrTiO3 interfaces. *Phys. Rev. B*, 67(153327), 2003.

[53] R. Raffaele. Macroscopic polarization in crystalline dielectrics: the geometric phase approach. *Rev. Mod. Phys.*, 66(3):899–915, Jul 1994.

[54] R.D. King-Smith and D. Vanderbilt. Theory of polarization of crystalline solids. *Phys. Rev. B*, 47(3):1651–1654, Jan 1993.

Chapter 7

Nanoscale Ferroelectric Films, Strips and Boxes

Jeffrey F. Webb

Additional information is available at the end of the chapter

1. Introduction

A major characteristic of ferroelectric materials is that they posses a non-zero polarization in the absence of any applied electric field provided that the temperature is below a certain critical temperature[1]; in the literature this is often called a spontaneous polarization, or an equilibrium polarization. Another important defining characteristic is that this polarization can be reversed by applying an electric field. This is the basis of a useful application of ferroelectrics in which binary information can be stored according to the polarization direction which can be switched between two states by an electric field. For example, a thin film of ferroelectric can be switched to different states at different regions by suitably patterned electrodes, thus creating a ferroelectric random access memory[2].

With the increasing miniaturization of devices it becomes important to investigate size-effects in ferroelectrics. Thin film geometries are of interest in which one spatial dimension is confined, as well as strip geometries in which two-dimensions are confined, and, what here we will call box geometry in which all three dimensions are confined. By confinement we mean that surfaces of the ferroelectric that intersect on at least one line going through the central region of the ferroelectric are sufficiently close to this region to make their presence cause a non-negligible effect on the ferroelectric as a whole so that the behaviour is different from what would be expected in a bulk region far from any surface. The geometries chosen are typical of the sort that can be fabricated in micro or nanoelectronics and fit conveniently into a Cartesian coordinate system.

The aim of this chapter is to show how the equilibrium polarization can be calculated in general for a confined volume of ferroelectric and this is applied to the three aforementioned geometries. In fact such calculations for thin films are well established[3–7]; here they are included with the other geometries, which are not so well studied, for completeness, and because they fit logical into the theoretical framework to be presented. A general problem of a similar nature to the one of interest here, but applied to spheres and cylinders has been considered by Morison's et al.[8]. This work employed approximate analytical solutions to the problem of calculating the polarization and avoided a full three-dimensional application of the general theory.

$T > T_C$ $T < T_C$

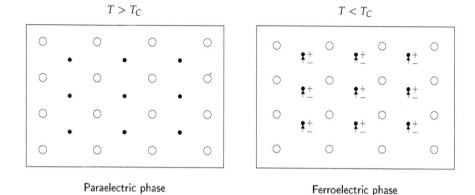

Paraelectric phase   Ferroelectric phase

Figure 1. A two-dimensional representation of how a crystal structure changes symmetry when passing between the paraelectric ($T > T_C$) and ferroelectric ($T < T_C$) phases

2. Foundations of the Landau-Devonshire theory of ferroelectrics

2.1. Bulk ferroelectrics

Since Landau-Devonshire theory is the theoretical basis for the work in this chapter it will be introduced here. Further details can be found elsewhere such as in Refs. [1, 9, 10]. In this section we consider the case of a bulk ferroelectric; the effect of surfaces will be introduced after that.

It is common to state that the starting point for Landau-Devonshire theory is the free energy per unit volume of a bulk ferroelectric expressed as an expansion in powers of the polarization P. The observed equilibrium polarization P_0 is then given by the value of P that minimizes the free energy. Here, however, following Strukov and Lenanyuk[10], we outline how this expansion comes from a statistical thermodynamic treatment involving an incomplete Gibbs potential, since this gives more insight into the theory. First a few more details about the nature of ferroelectrics.

When a ferroelectric crystal is above the critical temperature T_C there is no spontaneous polarization; this is the paraelectric or high symmetry phase in which the crystal structure is of higher symmetry than the ferroelectric phase below T_C since the appearance of the spontaneous polarization lowers the symmetry, as is evident from the two-dimensional representation in Fig. 1. Thus we see that during the transition from the paraelectric to ferroelectric state, a structural phase transition occurs. A consequence of this is that as the temperature approaches T_C the structure becomes easy to distort giving rise to anomalous peaks in some of the properties such as the dielectric function ϵ, as illustrated in Fig. 2. In fact, in the ferroelectric phase, a crystal may have different domains such that the polarization in adjacent domains is not in the same direction[9], however it is possible to produce single-domain crystals—by cooling through T_C in the presence of an electric field in one of the possible polarization directions for instance—and that is what we consider here.

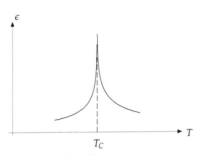

Figure 2. A peak in the dielectric function ϵ occurs as $T \to T_C$.

In the simplest case, then, a ferroelectric phase transition can be described by a single parameter that in Fig. 1 would be describable by the displacement of the sublattice from its central position. Since the paraelectric phase is of higher symmetry than the ferroelectric phase we can consider it to be more orderly and so the parameter is sometimes referred to as an order parameter. Actually the atoms are ions with charge and the corner-atom lattice would have a charge opposite to that of the interior sub-lattice. From the macroscopic point of view the corner atoms can be considered to produce a focus of charge at the center of each rectangle of opposite sign to the atoms that are displaced during the phase transition (see Fig. 1 where the centers of positive and negative charge are marked for the ferroelectric phase). For this reason the order parameter is often taken to be the polarization, which is the dipole moment per unit volume. For the time being, however, we will stay with the displacement representation of the order parameter as this is more directly related to the structural transition. If the displacement is denoted by η then it is clear that in the ferroelectric phase $\eta \neq 0$ and in the paraelectric phase $\eta = 0$.

The description above implies that each center ion is displaced in the same direction and this is the case for what are termed displacive ferroelectrics. There is another possibility, however, in which the displacement is not the same for every unit cell. For example, in some cells the displacement could be up while in others it could be down. Macroscopically ferroelectric and paraelectric phases can still result but now the ferroelectric phase occurs when statistically there are more displaced in one direction than in the other; and the paraelectric phase corresponds to the number of atoms displaced upwards is statistically equal to the number displaced downwards. In this case the ferroelectric is called an order-disorder ferroelectric. In terms of the macroscopic crystal symmetry however displacive and order-disorder ferroelectrics are equivalent and both can be described in general by Landau-Devonshire theory. The crystal structure of order-disorder ferroelectrics tends to be more complicated than for the displacive type[9, 10].

The basis of the Landau-Devonshire theory is that in the phase transition the ferroelectric phase may be represented by a distorted symmetrical phase. As outlined above this is a macroscopic view and the theory is phenomenological, as will be brought out further below. It turns out that the symmetry elements lost by the crystal at the transition temperature is sufficient information for a description of the anomalies of practically all of the thermodynamic properties of the crystal[10].

To make further progress consider the crystal as a system of N interacting particles with potential energy $U(\mathbf{r}_1, \ldots, \mathbf{r}_N)$. According to Gibbs[11, Chapter 8], the equilibrium thermodynamic potential at pressure p and temperature T associated with the potential energy of interaction of the particles is given by

$$\Phi(p, T) = -k_B T \ln Z, \tag{1}$$

where k_B is Boltzmann's constant and

$$Z = \int_{-\infty}^{\infty} \exp\left[-U(\mathbf{r}_1, \ldots, \mathbf{r}_N)/(k_B T)\right] \prod_i d\mathbf{r}_i. \tag{2}$$

The expression for Z in Eq. (2) comes from the probability that the value of the radius vector of the first particle lies between $\mathbf{r}_1$ and $\mathbf{r}_1 + d\mathbf{r}_1$ and similarly for the other vectors, being given by

$$dw = C \exp\left[-U(\mathbf{r}_1, \ldots, \mathbf{r}_N)/(k_B T)\right] \prod_i d\mathbf{r}_i. \tag{3}$$

Since integration of all of the variables must yield unity, $C = 1/Z$. The probability distribution can therefore be written as

$$dw = \exp\left\{[\Phi(p, T) - U(\mathbf{r}_1, \ldots, \mathbf{r}_N)]/(k_B T)\right\} \prod_i d\mathbf{r}_i. \tag{4}$$

A Gibbs thermodynamic potential for nonequilibrium states can also be formulated by using a larger number of variables than only p and T. The new variables are introduced via a linear transformation:

$$\mathbf{r}_1, \ldots, \mathbf{r}_N \rightarrow \xi_1, \ldots, \xi_{3N}. \tag{5}$$

We choose one of these variables to be η, which describes the nonequilibrium state of interest as explained above, by setting $\xi_1 = \eta$. Now

$$dw = \exp\left\{[\Phi(p, T) - U(\eta, \xi_2 \ldots, \xi_{3N})]/(k_B T)\right\} d\eta \, d\xi_2 \cdots \xi_{3N}, \tag{6}$$

the probability that the variables will lie in the ranges η to $\eta + d\eta$, ξ_2 to $\xi_2 + d\xi_2$, and so on. Focusing on the variable of interest η the probability of finding the system in a state in which this variable is in the range $\eta + d\eta$ is

$$dw(\eta) = d\eta \int_{-\infty}^{\infty} \exp\{[\Phi(p,T) - U(\eta,\xi_2,\ldots,\xi_{3N})] / (k_B T)\} d\xi_2 \cdots \xi_{3N}. \qquad (7)$$

The thermodynamic potential is now also a function of η, and we write

$$\Phi(p,T,\eta) = -k_B T \ln \int_{-\infty}^{\infty} \exp\{[-U(\eta,\xi_2,\ldots,\xi_{3N})] / (k_B T)\} d\xi_2 \cdots \xi_{3N}, \qquad (8)$$

and

$$dw(\eta) = \exp\{[\Phi(p,T) - \Phi(p,T,\eta)] / (k_B T)\} d\eta. \qquad (9)$$

Here it can be seen that the value of η that minimizes $\Phi(p,T,\eta)$ will maximize the distribution function and so this value corresponds to the equilibrium value of η. If $\Phi(p,T,\eta)$ is known, the usual equilibrium function $\Phi(p,T)$ can be found, using Eq. (9), from

$$\Phi(p,T) = -k_B T \ln \int_{-\infty}^{\infty} \exp\{[-\Phi(\eta)] / (k_B T)\} d\eta. \qquad (10)$$

Now, let the equilibrium value of η be η_0—which corresponds to the minimum value of $\Phi(p,T,\eta)$—and expand $\Phi(p,T,\eta)$ in a series about the point $\eta = \eta_0$:

$$\Phi(p,T,\eta) = \Phi(p,T,\eta_0) + \Delta\Phi(p,T,\eta - \eta_0) = \Phi(p,T,\eta_0) + \frac{1}{2}A(p,T)(\eta - \eta_0)^2 + \cdots. \qquad (11)$$

Using Eq. (10), $\Phi(p,T)$ can be written as

$$\Phi(p,T) = \Phi(p,T,\eta_0) - k_B T \ln \int_{-\infty}^{\infty} \exp\{[-\Delta\Phi(p,T,\eta - \eta_0)] / (k_B T)\} d\eta. \qquad (12)$$

In Eq. (12) the first term is the minimum of the potential as a function of p, T and η, and we are interested in the minimizing value $\eta = \eta_0$ for given values of p and T. The second term is the contribution to the thermodynamic potential of fluctuations in η with order of magnitude given by the thermal energy per degree of freedom $K_B T/2$. However since we are dealing with a large number of particles in a macroscopic sample ($\sim 10^{23}$) the fluctuations are very small and the second term can be neglected to give

$$\Phi(p,T) = \Phi(p,T,\eta_0). \qquad (13)$$

Note that if the total number of degrees of freedom (left unintegrated as the single degree η was) required to describe the system is comparable to N, then the neglect of the second term is no longer valid and Eq. (10) must be used in place of Eq. (13). We will not be concerned with such situations here, but very close to the transition temperature the Landau-Devonshire theory breaks down and the fluctuations must be considered, which can be done with the aid of Eq. (10), as discussed in Ref. [10].

In general thermodynamic functions which contain extra variables that remain unintegrated as η was above, are nonequilibrium thermodynamic functions known as incomplete functions. The general form is

$$\Phi(p, T, \eta_1, \eta_2, \ldots,, \eta_n) = \Phi_0(p, T) + \Phi_1(p, T, \eta_1, \eta_2, \ldots,, \eta_n). \tag{14}$$

The equilibrium values of the η_i are found by minimizing $\Phi(p, T, \eta_1, \eta_2, \ldots,, \eta_n)$ according to

$$\frac{\partial \Phi}{\partial \eta_1} = 0, \ldots, \frac{\partial \Phi}{\partial \eta_n} = 0, \tag{15}$$

which determine a set of minimizing values $\eta_{oi}(p, T)$. Substituting these values into Eq. (14) determines the equilibrium thermodynamic function

$$\Phi(p, T) = \Phi_0(p, T) + \Phi_1(p, T, \eta_{01}, \eta_{02}, \ldots,, \eta_{0n}). \tag{16}$$

This does not take fluctuations of the η_{0i} into account but the corresponding error will be small provided that $n \ll 3N$.

Although we are only considering a ferroelectric phase transition of the simplest type describable by introducing a single extra variable η into the thermodynamic potential the general form shows how to treat more complicated cases for which the polarization may occur in more ways than along one line, as we will see later. Polarization in three-dimensionss can be treated with three parameters, η_1 to η_3.

Returning to a one-component case we now show how the form of the thermodynamic potential can be worked out. As explained, in a ferroelectric phase transition we are dealing with displacements of certain atoms or groups of atoms; the structure of the nonsymmetric phase can be obtained from the structure of the symmetric phase by small displacements. Although the same lowering of symmetry in a phase transition can occur with different types of ordering implying that the choice of the order parameter is ambiguous, it turns out that the character of the anomalies of the physical properties in the transitions can be elucidated for any particular relationship between the order parameter and the displacements as long at the appearance of the order parameter leads to a symmetry change corresponding to that of the crystal's[10, 12].

In view of this we can take the order parameter as the extra variable of the incomplete thermodynamic potential. The phase transition can then be described at a given pressure if $\Phi(p, T, \eta)$ has a minimum at $\eta = 0$ in the symmetric phase ($T > T_C$), and at least two

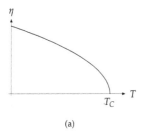
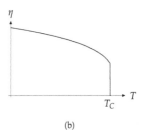

(a)   (b)

Figure 3. (a) Second order phase transition in which the order parameter decreases continuously to zero. (b) First order phase transition in which the order parameter displays a discontinuity. There is a continuous decrease in η before, which means that it is a first-order transition close to a second order transition. at $T = T_C$.

minima[1] at $\eta \neq 0$ in the nonsymmetric phase ($T < T_C$). The function $\Phi(p, T, \eta)$ must be consistent with this as T changes through the transition temperature so that it is determined for both the symmetric and nonsymmetric phases. These requirements place restrictions on the dependence of the potential Φ on η: Φ is a scalar function characterizing the physical properties of ferroelectric crystals, and as such must be invariant under any symmetry operations on the symmetric phase consistent with the crystal symmetry of this phase.

Before exploring further the form of Φ that describes the phase transition the difference between first and second order phase changes will be discussed. In a second order transition η decreases continuously to zero as the temperature is lowered through T_C; in a first order transition, on the other hand, there is a discontinuous jump of η to zero at $T = T_C$. Both situations are illustrated in Fig. 3. The first order transition shown in this figure is in fact close to a second order transition and so there is a continuous decrease in η before the jump and the jump is not very large. Crystals can exhibit first or second order transitions depending on the crystal structure[9, 10]. Either case can be treated by Landau-Devonshire theory, as long as the first order case is close to a second order transition. Turning back now to the development of an expression for Φ, we see that in the vicinity of a second order transition it is permissible to deal only with small lattice distortions, that is, small η. Hence the thermodynamic potential can be expanded into a series in η, with p and T treated as parameters, and we obtain

$$\Phi(p, T, \eta) = \Phi(p, T, 0) + \Phi'(p, T, 0)\eta + \frac{1}{2}\Phi'(p, T, 0)\eta^2 + \cdots . \qquad (17)$$

To represent a given crystal this expression must be invariant under symmetry operations that correspond to the symmetry elements of the crystal. A ferroelectric phase transition resulting in a spontaneous polarization occurs as long as the order parameter transforms as a vector component. Only crystal symmetries for which this is true are ferroelectric and then the order parameter can be considered to be equivalent to a component of the spontaneous polarization (this point is discussed in more detail in Refs.[9, 10, 12]).

---

[1] Part of the definition of a ferroelectric is that it has a reversible polarization in the nonsymmetric phase so that at least two minimum are required; otherwise the reverse direction would not be an equilibrium state.

$2/m$	1	2	m	$\bar{1}$
$\eta = P$	P	P	$-P$	$-P$

Table 1. Transformation of the order parameter

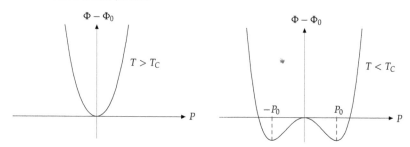

Figure 4. The form of $\Phi - \Phi_0$ that satisfies the basic requirements for a ferroelectric phase transition. When $T > T_C$ there is single minimum at $P = 0$ and there is no spontaneous polarization; for $T < T_C$ a spontaneous polarization can exist at one of the minima at $P = \pm P_0$.

To make further progress we consider a definite example of a symmetry which allows the order parameter to transform as a vector and so corresponds to a polarization having a single component; to reflect this correspondence we also make the notation change $\eta \to P$ where P is the polarization component.[2] Consider a second-order phase transition such that the change into the ferroelectric state at the critical temperature is accompanied by a symmetry change from the group $2/m$ (symmetry elements: 1, $\bar{1}$, m, 2) to 2 (symmetry elements: 1, 2). The ferroelectric triglycine sulphate shows this symmetry change during its phase transition[13, 14]. During the transition to the nonsymmetric phase the crystal loses the inversion center and the plane of symmetry, $\bar{1}$ and m. These disappearing symmetry elements correspond to the symmetry operations that reverse the sign of the order parameter $\eta = P$, as is shown in Table 1. Since Φ must be invariant under any transformations of the symmetrical phase it is clear from this that it cannot include terms linear in P or odd powers of P, since such terms would be changed by the operations m and $\bar{1}$.

Therefore the first $\eta = P$ dependent term in Eq. (17) will be of the form $A(p, T)P^2$. For simplicity we now assume that the pressure is at some fixed value and discuss the temperature dependence. The expression for Φ must reflect the fact that at $T > T_C$ there is a minimum at $P = 0$ so that the equilibrium state (represented by the minimum) is one in which $P = 0$, and when $T < T_C$ there will be two minima for which $P \neq 0$ so that the possible equilibrium states occur at nonzero values of P. This is satisfied if $A(T)$ passes continuously from $A(T < T_C) < 0$ to $A(T > T_C) > 0$ when $T > T_C$, with $A(T = T_C) = 0$, provided that the next allowable term $B(T)P^4$ (remembering that odd terms have the wrong symmetry) is included with $B(T) > 0$, as can be seen from Fig. 4. This results in a thermodynamic potential of the form

$$\Phi = \Phi_0 + A(T)P^2 + B(T)P^4. \tag{18}$$

---

[2] The general case in which the polarization has three components would be covered by three order parameters and the correspondence would be $(\eta_1 \to P_1, \eta_2 \to P_2, \eta_3 \to P_3)$, with $\mathbf{P} = (P_1, P_2, P_3)$.

Generally the exact form of the temperature dependence may be difficult to find. However near the phase transition temperature a series expansion in powers of $T - T_C$ (at a given pressure) can be used to give

$$A(T) = A(T_C) + A_1(T - T_C)A_2(T - T_C)^2 + \cdots \tag{19}$$

$$B(T) = B(T_C) + B_1(T - T_C)B_2(T - T_C)^2 + \cdots . \tag{20}$$

Taking into account the above mentioned properties: $A(T_C) = 0$ and $B(T_C) > 0$, the simplest forms retaining the essential non-zero first terms gives

$$A(T) = A_1(T - T_C) = \frac{1}{2}a(T - T_C) \tag{21}$$

$$B(T) = B(T_C) = \frac{1}{4}b. \tag{22}$$

The numerical factors are introduced for convenience, as will be clear shortly. We thus obtain, for a given pressure,

$$\Phi(T, P) = \Phi_0(T) + \frac{1}{2}a(T - T_C)P^2 + \frac{1}{4}bP^4. \tag{23}$$

Usually it is sufficient to consider the energy due to the presence of the spontaneous polarization per unit volume (we are dealing with a bulk sample assumed uniform of the crystal volume) so that it is convenient to write, for a volume v of ferroelectric,

$$F = \frac{\Phi(T, P) - \Phi_0(T)}{v} = \frac{1}{2}\alpha(T - T_C)P^2 + \frac{1}{4}\beta P^4, \tag{24}$$

where $\alpha = a/v > 0$, $\beta = b/v > 0$, and we refer to F as the Gibbs free energy density.

The equilibrium polarization P_0 is now found by minimizing F from the conditions

$$\frac{\partial F}{\partial P} = 0, \quad \text{and} \quad \frac{\partial^2 F}{\partial P^2} > 0.$$

From this we find

$$P_0^2 = \begin{cases} 0 & \text{for } T > T_C, \\ -\frac{\alpha(T - T_C)}{\beta} & \text{for } T < T_C. \end{cases} \tag{25}$$

Here we see that it is necessary to differentiate to find P_0 which explains why the numerical coefficients $1/2$ and $1/4$ were introduced into the expansion of the thermodynamic potential.

The form of F in Eq. (24) is sufficient to describe a second order phase transition since $P_0 \to 0$ continuously as $T \to T_C$. For second order transitions in which there is a discontinuous jump, Eq. (24) is insufficient. However a first-order transitions near to a second-order transition the

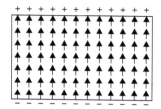

Figure 5. A component of polarization that is directed along a normal to the plane of the film when $T < T_C$. The resulting surface charges with opposite sides having signs of opposite charge, creates an electric field which in the opposite direction to the polarization.

phase change can be described by adding a further term in P^6 (a term in P^5, since it is an odd power, does not fulfill the symmetry requirements) which results in an energy density given by

$$F = \frac{1}{2}\alpha(T - T_C)P^2 + \frac{1}{4}\beta P^4 + \frac{1}{6}\gamma P^6. \tag{26}$$

But, whereas $\beta > 0$ for a second order transition, for F to exhibit a discontinuous jump it is necessary to choose $\beta < 0$ with $\gamma > 0$. More on this can be found in Lines and Glass[9].

We have thus shown in some detail how it is that a relatively simple free energy expression can account for the main characteristics of a ferroelectric, with the observed equilibrium polarization found by minimizing the free energy. Other terms can be added to the free energy to account for various external influences. For example an external electric field with component E along the direction of polarization can be accounted by first substituting the equilibrium polarization P_0 into F and adding a term EP_0. The corresponding susceptibility is then given by $\chi = \frac{\partial^2 F}{\partial E^2}$.

Note that Landau-Devonshire theory is a phenomenological theory of the macroscopic properties of the ferroelectric: the coefficients, α, β, etc., are not derivable from it; instead they must be found from experiment, or, in some cases, can be found from first-principle calculations[1].

2.2. Extension to thin films

Here a free-standing thin film is considered. If the equilibrium polarization below T_C has a component that is aligned with a normal to the plane of the film, then depolarization effects due to the space charge that appears at the surfaces, as illustrated in Fig. 5, must be taken into account. At first we avoid this complication by assuming a polarization that is in-plane. Also to remain with a one dimensional treatment for now we assume that this this is along the x direction of a Cartesian coordinate system and that the plane surfaces of the film are normal to the z axis at $z = 0$ and $z = L$, where L is the thickness of the film. The effect of the surfaces is such that the polarization near them may differ from the bulk value. For the case just described this implies that $P = P(z)$. It can be shown[1, 3, 15] that this can be accounted for in the free energy by adding surface terms involving P^2 integrated over the surfaces, and a gradient term $|dp/dz|$. The free energy is now given by

$$G = \left(\iint dx dy \right) \int_0^L f(P, dP/dz)\, dz + \frac{1}{2} \frac{D}{d} \left(\iint dx\, dy \right) \left[P^2(0) + P^2(L) \right], \qquad (27)$$

in which

$$f(P, dP/dz) = \frac{1}{2} A P^2 + \frac{1}{4} B P^4 + \frac{1}{6} C P^6 + \frac{1}{2} D \left(\frac{dP}{dz} \right)^2, \qquad (28)$$

the temperature dependence is in A through $A = a(T - T_C)$, and the integrals over x and y are factored out of the three dimensional integral over a volume of the film because P does not vary in these directions. Thus any surface area can be chosen to integrate over so that, if this areas is S, then $\iint dx dy = S$, and we can write a free energy per unit area as

$$F = \frac{G}{S} = \int_0^L dz \left[\frac{1}{2} A P^2 + \frac{1}{4} B P^4 + \frac{1}{6} C P^6 + \frac{1}{2} D \left(\frac{dP}{dz} \right)^2 \right] + \frac{1}{2} D \left[\frac{P^2(0)}{d} + \frac{P^2(L)}{d} \right]. \qquad (29)$$

Here since the film is free standing with identical surface properties for both surfaces the surface terms at $z = 0$ and $z = L$ involve the same factor, $1/d$, which simplifies the problem. If the film was not free standing, with for example, one surface an interface with a substrate, then this could be modeled by introducing different factors[6] so that the last term in square brackets in Eq. (29) would be replaced by $P(0)/d_1 + P(L)/d_2$.

The equilibrium polarization is still the P that minimizes F, but now the problem is to find the function $P(z) = P_0(z)$ that minimizes F rather than a finite set of values. So we are dealing with the minimization of a functional and the classical methods from the calculus of variations[16, Chapter 4] may be used, which cast the problem into the form of a differential equation known as the Euler-Lagrange equation and the function that minimizes F is the solution to this equation for a set of boundary conditions that also need to be specified. In this case, as will be shown below the boundary conditions used are implicit in Eq. (29) and are equivalent to allowing the polarization to be free at the film surfaces rather than fixed. More details on the method for the general three dimensional case are given in Section 4. Here we simply state that the classical methods lead to the following Euler-Lagrange equation and boundary conditions for the equilibrium polarization in the film:

$$D \frac{d^2 P_0}{dz^2} - A P_0 - B P_0^3 - C P_0^5 = 0. \qquad (30)$$

With boundary conditions

$$\frac{dP_0}{dz} - \frac{1}{d} P_0 = 0, \quad \text{at } z = 0, \qquad (31)$$

and

$$\frac{dP_0}{dz} + \frac{1}{d} P_0 = 0, \quad \text{at } z = L. \qquad (32)$$

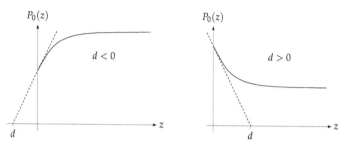

Figure 6. The extrapolation length d for one of the film surfaces at $z = 0$. For $d < 0$ the polarization turns down as the surface is approached; for $d > 0$ it turns up; this is also true for the surface at $z = L$ (not shown). Here the extrapolated gradient line crosses the z-axis at $z = 0 + d$; for the surface at $z = L$ it would cross at $z = L + d$.

It can be seen from the boundary conditions that d can be interpreted as an extrapolation length, as illustrated in Fig. 6. For $d < 0$ the polarization turns upwards as it approaches a surface; for $d > 0$ it turns down as a surface is approached.

Now the task is to solve the differential equation—Eq. (30) which is nonlinear—subject to the boundary conditions. For first order transitions in which $C \neq 0$ the equation must be solved numerically[17, 18]. However, for the simpler case of second-order transitions ($C = 0$) even though the equation is still nonlinear, an analytical solution can be found in terms of elliptic functions[3, 19, 20] and is given by

$$P_0(z) = P_1 \, \text{sn} \left[K(\lambda) - \frac{z - L/2}{\zeta}, \lambda \right], \quad \text{for } d < 0, \tag{33}$$

where sn is the Jacobian elliptic function with modulus $\lambda = P_1/P_2$, $K(\lambda)$ is the complete elliptic integral of the first kind (see Ref. [21] for more on elliptic functions and integrals),

$$P_1^2 = -\frac{A}{B} - \sqrt{\frac{A^2}{B^2} - \frac{4G}{B}}, \tag{34}$$

$$P_2^2 = -\frac{A}{B} + \sqrt{\frac{A^2}{B^2} - \frac{4G}{B}}, \tag{35}$$

$$\zeta = \frac{1}{P_2} \sqrt{\frac{2D}{B}}, \tag{36}$$

and G is found by substituting it into the boundary conditions and solving the resulting transcendental equations numerically.

A plot of the polarization profile according to Eq. (33) is given in Fig. 7 for parameter values given in the figure caption. The polarization throughout the film decreases continuously to zero as the temperature increases and approaches T_C. So there are two states: a ferroelectric state below T_C for which there is a polarization everywhere in the film and a paraelectric state above T_C for which the polarization is zero everywhere.

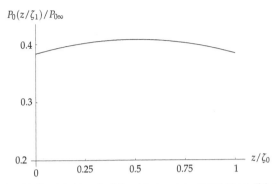

Figure 7. Polarization in a thin film plotted from Eq. (33) and the boundary conditions in Eqs. (31) and (32). Dimensionless variables and parameters used are: $P_{0\infty} = (aT_C/B)^{1/2}$ (the spontaneous polarization for a bulk ferroelectric), $\zeta_0 = [2D/(aT_C)]^{1/2}$, $\zeta_1 = P_{0\infty}/P_2$ (P_2 is given by Eq. (35)), $\Delta T' = (T - T_C)/T_C = -0.4$, $L' = L/\zeta_0 = 1$, $d' = 4L'$, $G' = 4GB/(a/T_C)^2 = 0.105$ (found from the boundary conditions).

The case $d > 0$ is complicated by the appearance of a surface term between T_C and a lower temperature T_{C0} for which the regions near the film surfaces have a spontaneous polarization but the polarization in the interior is still zero. Only below T_{C0} is there a nonzero polarization in the interior of the film. The form of the solution, although still expressible in terms of an elliptic function is somewhat different for the surface state, as is discussed elsewhere[3, 19, 20]

The above was for in-plane polarization which, as has been explained, avoids the need to consider a depolarization field induced by the polarization. In general, however, there may be an out-of-plane component that would give rise to a depolarization field. Theoretical descriptions of the depolarization field have been considered by several authors[5, 7, 22, 23]. For the case in which there are no free charges at the surfaces it can be shown[5, 22] from Maxwell's equations that a term

$$\frac{1}{2\epsilon_0\epsilon_\infty}\left[P(P - \langle P\rangle)\right] \tag{37}$$

added to the energy density in Eq. (28) accounts for the effect of the depolarization field provided that P is now taken to be directed along a normal to the plane of the the film (as it is in Fig. 5). In Eq. (37) ϵ_0 is the permittivity of free space and ϵ_∞ is the background dielectric constant of the film[5, 24], and

$$\langle P\rangle = \frac{1}{L}\int_0^L P\,dz, \tag{38}$$

which is the average value of $P(z)$ in the film.

3. The free energy density appropriate for a three-dimensional treatment of a ferroelectric crystal

The general form of the free energy density for the interior of a ferroelectric crystal in three-dimensions can be written as series expansion in components of the polarization, together with terms taking into account the influence of surfaces and depolarization fields. The region occupied by the interior of the ferroelectric (the surface energy will appear later as a surface energy density integrated over the surface), is taken to be the interior of a rectangular box defined by

$$V = \left\{ \mathbf{x} \mid x_i \in \left(l_i^-, l_i^+ \right), i = 1, 2, 3 \right\}, \tag{39}$$

so that the sides of the box are given by $l_i = l_i^+ - l_i^-$, $i = 1, 2, 3$. Extending the formalism in the previous sections to three dimensions, the energy density can be written

$$f\left(\mathbf{x}, \mathbf{P}, \mathbf{P}_{x_1}, \mathbf{P}_{x_2}, \mathbf{P}_{x_3}, \langle \mathbf{P} \rangle \right) = f_{\text{series}}\left(\mathbf{x}, \mathbf{P} \right) + f_{\text{grad}}\left(\mathbf{x}, \mathbf{P}_{x_1}, \mathbf{P}_{x_2}, \mathbf{P}_{x_3} \right) + f_{\text{dep}}\left(\mathbf{x}, \mathbf{P}, \langle \mathbf{P} \rangle \right) \tag{40}$$

The notation used for the arguments, which will be useful when applying calculus of variations methods in the next section, is as follows: $\mathbf{x} = (x_1, x_2, x_3)$, for a position in the crystal in Cartesian coordinates; $\mathbf{P} = (P_1, P_2, P_3)$, the polarization vector; $P_i = P_i(x_1, x_2, x_3)$, $i = 1, 2, 3$; $\mathbf{P}_{x_i} = \partial \mathbf{P}/\partial x_i$, $i = 1, 2, 3$, and the averages vector is $\langle \mathbf{P} \rangle = \left(\langle P_1 \rangle, \langle P_2 \rangle, \langle P_3 \rangle \right)$ in which

$$\langle P_i \rangle = \frac{1}{l_i} \int_{l_i^-}^{l_i^+} P_i \, dx_i. \tag{41}$$

The terms on the right of Eq. (40) are given next. The first of these terms f_{series} is the series expansion part of the free energy (due to the polarization, without a constant first term) in terms of the components of the polarization. A general series expansion in these terms can be written as

$$f_{\text{series}}\left(\mathbf{x}, \mathbf{P} \right) = \sum_{i=1}^{\infty} \frac{1}{i} \left(\sum_{\alpha_1 \geqslant \cdots \geqslant \alpha_i} A_{\alpha_1 \cdots \alpha_i} P_{\alpha_1} \cdots P_{\alpha_i} \right). \tag{42}$$

Here each α_i runs from x_1 to x_3 and the notation $\alpha_1 \cdots \alpha_i$ indicates that all permutations of $P_{\alpha_1} \cdots P_{\alpha_i}$ of the ith term, which would otherwise appear as separate, have already been summed: the inequalities insure that only one permutation is present for the ith term with coefficient $A_{\alpha_1 \cdots \alpha_i}$. However in practice, depending on the crystal symmetry, terms which are not separately invariant under the symmetry transformations of the symmetry group of the crystal, will not appear.

The term f_{grad}, a gradient term, is given by

$$f_{\text{grad}}\left(\mathbf{x}, \mathbf{P}_{x_1}, \mathbf{P}_{x_2}, \mathbf{P}_{x_3} \right) = \frac{1}{2}\delta \left[|\nabla P_1|^2 + |\nabla P_2|^2 + |\nabla P_3|^2 \right] = \frac{1}{2}\delta \left[\left| \frac{\partial \mathbf{P}}{\partial x_1} \right|^2 + \left| \frac{\partial \mathbf{P}}{\partial x_2} \right|^2 + \left| \frac{\partial \mathbf{P}}{\partial x_3} \right|^2 \right]. \tag{43}$$

It is easy to show that $\frac{1}{2}\delta \left[|\nabla P_1|^2 + |\nabla P_2|^2 + |\nabla P_3|^2 \right]$ is equal to the right hand side of Eq. (43). Both forms are shown so that it can be seen that f_{grad} involves gradient terms. However the other form is sometimes convenient to work with. Note that in general there are other symmetry allowed terms such as $\left[\left| \frac{\partial \mathbf{P}}{\partial x_1} \right|^2 \left| \frac{\partial \mathbf{P}}{\partial x_2} \right|^2 \right]$. However usually such terms are ignored

– a practice which will be followed here. Nonetheless, in the future it may be interesting to study such terms to extended the formalism given here.

Finally f_{dep} is the term due to the depolarization field for the general case in which the polarization has components along x_1, x_2 and x_3. It is given by

$$f_{dep}\left(\mathbf{x},\mathbf{P},\langle\mathbf{P}\rangle\right) = \frac{1}{2\epsilon_0\epsilon_\infty}\left[P_1\left(P_1 - \langle P_1\rangle\right) + P_2\left(P_2 - \langle P_2\rangle\right) + P_3\left(P_3 - \langle P_3\rangle\right)\right]. \qquad (44)$$

Having dealt with the energy density for the interior of the ferroelectric the remaining energy density is the surface energy density at any point on the surface of the ferroelectric. This will be proportional to P^2, where $P = |\mathbf{P}|$ and, as will be evident below, it is convenient to write it as

$$f_{surf}\left(\mathbf{x},\mathbf{P}\right) = \frac{\delta}{2d(\mathbf{x})}P^2(\mathbf{x}) \quad \text{for} \quad \mathbf{x} \in S, \qquad (45)$$

where S is the entire surface of the ferroelectric box given by

$$S = S_{x_1=l_1^-} \cup S_{x_1=l_1^+} \cup S_{x_2=l_2^-} \cup S_{x_2=l_2^+} \cup S_{x_3=l_3^-} \cup S_{x_3=l_3^+} \qquad (46)$$

where the sides of the box at $x_i = l_i^\mp$ are given by

$$S_{x_j=l_j^\mp} = \left\{\mathbf{x} \mid x_i = l_i^\mp, x_{\sigma^j(i)} \in \left[l_i^-, l_i^+\right], j = 1, 2\right\}, \qquad (47)$$

where we have introduced the cyclic operator σ which performs the operation $x_1 \mapsto x_2 \mapsto x_3 \mapsto x_1$, so that

$$\sigma(x_i) = x_{j+1} \pmod{3}; \qquad (48)$$

also σ applied n times where $n \geqslant 0$ is denoted by σ^n, and the definition of the operator is extended to when the x_i are arguments of a function h such that

$$\sigma\left(h(x_1,\ldots,x_3)\right) = h\left(\sigma(x_1),\ldots,\sigma(x_3)\right). \qquad (49)$$

It is easy to see that many of the terms in the free energy can be written in a shortened form with this notation because successive terms can often be generated by cyclic permutations of a starting term; later this operator notation will be useful when the free energy for the box is inserted into the Euler-Lagrange equations in Section 5.

Showing the form of the free energy densities for the ferroelectric box helps pave a way for dealing with the general problem of finding the minimum of the free energy density for an arbitrary volume of ferroelectric, which will be discussed next.

4. The calculus of variations applied to the polarization of a ferroelectric of arbitrary shape in three dimensions

The next step towards working out the equilibrium polarization for ferroelectric boxes and strips—an extension of the theory for thin films—is to work out in general the free energy for an arbitrary volume of ferroelectric. The minimization of this, as for the thin film, involves the minimization of a functional, which involves the calculus of variations and will be dealt with using classical methods.

It can be seen from the previous section that the general form of the free energy for a ferroelectric in the region V in $\mathbb{R}^3$ bounded by the closed surface S is given by

$$G = \int_V f\left(\mathbf{x}, \mathbf{P}, \mathbf{P}_{x_1}, \mathbf{P}_{x_2}, \mathbf{P}_{x_3}, \langle \mathbf{P} \rangle\right) dV + \int_S f_{\text{surf}}\left(\mathbf{x}, \mathbf{P}\right) dS. \tag{50}$$

Here G depends on whatever function $\mathbf{P}$ is, so the domain of G is a function space and the equilibrium polarization distribution is that function $\mathbf{P}_0$ in the space which minimizes G. At this point certain boundary conditions could be imposed on $\mathbf{P}$ according to the physical situation. For a ferroelectric crystal no such conditions are imposed which implies that $\mathbf{P}$ is free at the boundaries. However natural boundary conditions (that do not fix $\mathbf{P}$) will emerge from the minimization as we will see.

The minimization of the functional G in Eq. (50) will be carried out using a classical technique due to Euler[16] which involves considering a variation around $\mathbf{P}_0 = (P_{01}, P_{02}, P_{03})$ in function space. To this end we write

$$\mathbf{P}(\mathbf{x}) = \mathbf{P}_0(\mathbf{x}) + \epsilon \boldsymbol{\eta}(\mathbf{x}), \tag{51}$$

in which ϵ is a variable parameter and $\boldsymbol{\eta} = (\eta_1, \eta_2, \eta_3)$ is an arbitrary vector function. G can now be expressed as

$$G(\epsilon) = \int_V f\left(\mathbf{x}, \mathbf{P}_0 + \epsilon \boldsymbol{\eta}, \mathbf{P}_{0x_1} + \epsilon \boldsymbol{\eta}_{x_1}, \mathbf{P}_{0x_2} + \epsilon \boldsymbol{\eta}_{x_2}, \mathbf{P}_{0x_3} + \epsilon \boldsymbol{\eta}_{x_3}, \langle \mathbf{P}_0 + \epsilon \boldsymbol{\eta} \rangle\right), \tag{52}$$

where

$$\mathbf{P}_{0x_i} = \frac{\partial \mathbf{P}_0}{\partial x_i}. \tag{53}$$

The key idea for finding the minimizing function $\mathbf{P}_0$ is that a necessary condition for its existence is

$$\left.\frac{dG}{d\epsilon}\right|_{\epsilon=0} = 0.$$

Applying this condition to Eq. (52), and using the chain rule when differentiating the integrands, we find

$$\left.\frac{dG}{d\epsilon}\right|_{\epsilon=0} = \int_V \{(\nabla_{\mathbf{P}_0} f) \cdot \boldsymbol{\eta} + \left(\nabla_{\mathbf{P}_{0x_1}} f\right) \cdot \boldsymbol{\eta}_{x_1} + \left(\nabla_{\mathbf{P}_{0x_2}} f\right) \cdot \boldsymbol{\eta}_{x_2} + \left(\nabla_{\mathbf{P}_{0x_3}} f\right) \cdot \boldsymbol{\eta}_{x_3}$$

$$+ \left(\nabla_{\langle \mathbf{P}_0 \rangle} f\right) \cdot \langle \boldsymbol{\eta} \rangle \} \, dV + \int_S (\nabla_{\mathbf{P}_0} f_{\text{surf}}) \cdot \boldsymbol{\eta} \, dS = 0, \quad (54)$$

in which the notation

$$(\nabla_{\mathbf{A}} f) \cdot \mathbf{B} = \frac{\partial f}{\partial A_1} B_1 + \frac{\partial f}{\partial A_2} B_2 + \frac{\partial f}{\partial A_3} B_3 \quad (55)$$

has been introduced.

If all terms in the integrands in Eq. (54) could be written as a dot product with $\boldsymbol{\eta}$, as is the case for the first and last terms on the right of Eq. (54), then $\boldsymbol{\eta}$ would appear as a single factor in the volume and surface integrands. Eq. (54) could then be written in the form $\int_V \boldsymbol{\eta} \cdot \Lambda_1 + \int_S \boldsymbol{\eta} \cdot \Lambda_2$, where Λ_1 and Λ_2 do not involve $\boldsymbol{\eta}$. As shown below, the Euler-Lagrange equations and boundary conditions will then follow by setting $\Lambda_1 = \Lambda_2 = 0$, which will satisfy Eq. (54). Therefore we now proceed to express Eq. (54) in the $\boldsymbol{\eta}$ factored form.

To facilitate this we start by writing Eq. (54) as

$$\left.\frac{dG}{d\epsilon}\right|_{\epsilon=0} = \int_V \{(\nabla_{\mathbf{P}_0} f) \cdot \boldsymbol{\eta} + \mathcal{I}_1 + \mathcal{I}_2 \} \, dV + \int_S (\nabla_{\mathbf{P}_0} f_{\text{surf}}) \cdot \boldsymbol{\eta} \, dS = 0, \quad (56)$$

where

$$\mathcal{I}_1 = \left(\nabla_{\mathbf{P}_{0x_1}} f\right) \cdot \boldsymbol{\eta}_{x_1} + \left(\nabla_{\mathbf{P}_{0x_2}} f\right) \cdot \boldsymbol{\eta}_{x_2} + \left(\nabla_{\mathbf{P}_{0x_3}} f\right) \cdot \boldsymbol{\eta}_{x_3}, \quad (57)$$

$$\mathcal{I}_2 = \left(\nabla_{\langle \mathbf{P}_0 \rangle} f\right) \cdot \langle \boldsymbol{\eta} \rangle. \quad (58)$$

Now we want to find Λ_1 and Λ_2 such that $\mathcal{I}_1 = \boldsymbol{\eta} \cdot \Lambda_1$ and $\mathcal{I}_2 = \boldsymbol{\eta} \cdot \Lambda_2$.

Starting with $\mathcal{I}_1$, by expanding the dot products and re-grouping the terms, it is easy to show that it can be rewritten as

$$\mathcal{I}_1 = (\nabla_{P_{01}x} f) \cdot (\nabla \eta_1) + (\nabla_{P_{02}x} f) \cdot (\nabla \eta_2) + (\nabla_{P_{03}x} f) \cdot (\nabla \eta_3), \quad (59)$$

in which the following notation has been used:

$$\nabla \mathbf{A} = \left(\frac{\partial A_1}{\partial x_1}, \frac{\partial A_2}{\partial x_2}, \frac{\partial A_3}{\partial x_3}\right) = \frac{\partial A_1}{\partial x_1} \hat{\mathbf{x}}_1 + \frac{\partial A_2}{\partial x_2} \hat{\mathbf{x}}_2 + \frac{\partial A_3}{\partial x_3} \hat{\mathbf{x}}_3, \quad (60)$$

the normal gradient operation, where $\hat{\mathbf{x}}_i$, $i = 1, 2, 3$, are unit vectors along the corresponding x_i axes of a Cartesian coordinate system. Also,

$$\nabla_{A_{ix}} = \left(\frac{\partial}{\partial A_{ix_1}}, \frac{\partial}{\partial A_{ix_2}}, \frac{\partial}{\partial A_{ix_3}} \right),$$

(61)

with $A_{ix_j} = \partial A_i / \partial x_j$.

Now, we utilize the vector identity

$$(\nabla \phi) \cdot \mathbf{A} \equiv \nabla \cdot (\phi \mathbf{A}) - \phi (\nabla \cdot \mathbf{A}),$$

(62)

together with the divergence theorem

$$\int_V \nabla \cdot \mathbf{A} \, dV = \oint_S \mathbf{A} \cdot (\hat{\mathbf{n}} \, dS),$$

(63)

where at each point on the surface S, $\hat{\mathbf{n}} = \hat{\mathbf{n}}(\mathbf{x}|_S)$ is a unit normal vector directed outwards (from the enclosed volume). Using Eqs. (62) and (63), it follows that

$$\int_V \mathcal{I}_1 \, dV = -\int_V \boldsymbol{\eta} \cdot \left[\sum_{i=1}^{3} \nabla \cdot (\nabla_{P_{0i}x} f) \, \hat{\mathbf{x}}_i \right] dV + \int_S \boldsymbol{\eta} \cdot \left[\sum_{i=1}^{3} \left\{ (\nabla_{P_{0i}x} f) \cdot \hat{\mathbf{n}} \right\} \hat{\mathbf{x}}_i \right] dS.$$

(64)

Next the $\mathcal{I}_2$ term is dealt with. But here is is better to look at it together with the integral from the outset, by considering

$$\int_V \mathcal{I}_2 \, dV = \int_V \left(\nabla_{\langle P_0 \rangle} f \right) \cdot \langle \boldsymbol{\eta} \rangle \, dV = \int_V \left(\sum_{i=1}^{3} \frac{\partial f}{\partial \langle P_i \rangle} \langle \eta_1 \rangle \right) dV.$$

(65)

The reason for considering $\mathcal{I}_2$ inside the integral is that, remembering that $\langle \eta_i \rangle$ is proportional to $\int_{l_i^-}^{l_i^+} \eta_i \, dx_i$, Eq. (65) can be written as

$$\int_V \mathcal{I}_2 \, dV = \int_V \left(\sum_{i=1}^{3} \eta_i \left\langle \frac{\partial f}{\partial P_{0i}} \right\rangle \right) dV = \int_V \boldsymbol{\eta} \cdot \left\langle \nabla_{\langle P_0 \rangle} f \right\rangle dV,$$

(66)

by changing the order of integration. In doing this we have used the fact that $dV \equiv dx_1 dx_2 dx_3$ and $\int_V \equiv \int_{x_1} \int_{x_2} \int_{x_3}$.

The goal of factoring out $\boldsymbol{\eta}$ involving a dot product has now been achieved and Eq. (54) can be written, using Eqs. (64) and (66), as

$$\int_V \boldsymbol{\eta} \cdot \left[\boldsymbol{\nabla}_{\mathbf{P}_0} f - \sum_{i=1}^{3} \boldsymbol{\nabla} \cdot \left(\boldsymbol{\nabla}_{P_{0i}\mathbf{x}} f \right) \hat{\mathbf{x}}_i + \left\langle \boldsymbol{\nabla}_{\langle \mathbf{P}_0 \rangle} f \right\rangle \right] dV$$

$$+ \int_S \boldsymbol{\eta} \cdot \left[\sum_{i=1}^{3} \left\{ \left(\boldsymbol{\nabla}_{P_{0i}\mathbf{x}} f \right) \cdot \hat{\mathbf{n}} \right\} \hat{\mathbf{x}}_i + \boldsymbol{\nabla}_{\mathbf{P}_0} f_{\text{surf}} \right] dS = 0. \quad (67)$$

However, at first we consider a restricted function space such that $\boldsymbol{\eta}$ is zero on S; then the surface integral in Eq. (67) vanishes resulting in

$$\int_V \boldsymbol{\eta} \cdot \left[\boldsymbol{\nabla}_{\mathbf{P}_0} f - \sum_{i=1}^{3} \boldsymbol{\nabla} \cdot \left(\boldsymbol{\nabla}_{P_{0i}\mathbf{x}} f \right) \hat{\mathbf{x}}_i + \left\langle \boldsymbol{\nabla}_{\langle \mathbf{P}_0 \rangle} f \right\rangle \right] dV = 0. \quad (68)$$

For this to be zero for any $\boldsymbol{\eta}$ in the restricted space, it follows from the lemma of du Bois-Reymond[25], that

$$\boldsymbol{\nabla}_{\mathbf{P}_0} f - \sum_{i=1}^{3} \boldsymbol{\nabla} \cdot \left(\boldsymbol{\nabla}_{P_{0i}\mathbf{x}} f \right) \hat{\mathbf{x}}_i + \left\langle \boldsymbol{\nabla}_{\langle \mathbf{P}_0 \rangle} f \right\rangle = 0, \quad (69)$$

which can be rewritten as

$$\frac{\partial f}{\partial P_{0i}} - \boldsymbol{\nabla} \cdot \left(\boldsymbol{\nabla}_{P_{0i}\mathbf{x}} f \right) + \left\langle \frac{\partial f}{\partial \langle P_{0i} \rangle} \right\rangle = 0, \quad i = 1, 2, 3. \quad (70)$$

These form the Euler-Lagrange equations that we seek. It is clear that the lifting of the restriction on $\boldsymbol{\eta}$ so that it is not necessarily zero on S, means that the Euler-Lagrange equations still result from Eq. (67), provided the lemma of du Bois-Reymond is applied to the surface integral term. This implies the boundary conditions

$$\left(\boldsymbol{\nabla}_{P_{ix}} f \right) \cdot \hat{\mathbf{n}} + \frac{\partial f_{\text{surf}}}{\partial P_i} \quad \text{on } S, \quad i = 1, 2, 3. \quad (71)$$

For any i the first term is a natural boundary condition arising from the minimization (not a prescribed condition), and the second term is due to the surface energy.

Thus we have shown that the equilibrium polarization for a ferroelectric of arbitrary shape in the region V is found by solving the Euler-Lagrange equations in Eq. (70), subject to the boundary conditions in Eq. (71). This formulation can be applied to finding the polarization in ferroelectrics with the geometries, box, strip and film, as will be shown in the next section for the case of a ferroelectric exhibiting a symmetric phase of cubic symmetry.

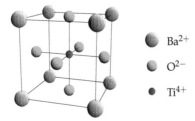

Figure 8. A structural unit for barium titanate, BaTiO$_3$.

5. The equilibrium Polarization in ferroelectric boxes, strips and films for cubic symmetry

Cubic symmetry is chosen because it is the symmetry of the popular perovskite ferroelectrics such as barium titanate (BaTiO$_3$) and lead titanate (PbTiO$_3$). A structural unit for barium titanate is represented in Fig. 8

The interest is in nanoscale boxes, strips and films. The formalism itself could easily be applied to thicker films, however for confinement in the nanoscale (in all three dimensions for boxes, in two dimensions in strips and in one for films) it is expected that the influence of the change in the polarization as it approaches a surface will have a more significant effect that at lager scales where the bulk properties would be more dominant, and so is sometimes referred to as a size effect.

5.1. Polarization for a ferroelectric nano-box

For a cubic crystal only those terms in f given by Eq. (40) that are invariant under the symmetry 16 operations of the cubic symmetry group[26] are allowed.[3] It can be shown[10, 12] that these are given by

$$f_{\text{cubic}}\left(\mathbf{x}, \mathbf{P}, \mathbf{P}_{x_1}, \mathbf{P}_{x_2}, \mathbf{P}_{x_3}, \langle \mathbf{P} \rangle\right) = \alpha_1 (T - T_C) P^2 + \sum_{i=2}^{3} \frac{1}{2i} \alpha_i \left(P^2\right)^i$$

$$+ \frac{1}{2} \beta_2 \sum_{i=0}^{2} \sigma^i \left(P_1^2 P_2^2\right) + \frac{1}{2} \beta_3 \sum_{i=0}^{2} \sigma^i \left(P_1^4 \left(P_2^2 + P_3^2\right)\right) + \frac{1}{2} \gamma_3 P_1^2 P_2^2 P_3^2$$

$$+ \frac{1}{2} \delta \sum_{i=1}^{3} \left| \frac{\partial \mathbf{P}}{\partial x_i} \right|^2 + \frac{1}{2 \epsilon_0 \epsilon_\infty} \sum_{i=1}^{3} P_i \left(P_i - \langle P_i \rangle\right), \quad (72)$$

where the cyclic operator defined by Eqs. (48) and (49) is being used and, to lighten the notation a little we make the replacement $\mathbf{P}_0 \to \mathbf{P} \Rightarrow P_{oi} \to P_i$, although the subscript that distinguishes the equilibrium polarization from the general function space remains implied.

---

[3] The operations are: $1, \bar{1}, 4, 4^{-1}, \bar{4}, \bar{4}^{-1}, 2', 2'', 2_x, 2_y, 2_z, m_1, m_2, m_3, m_4, m_5$.

The Euler-Lagrange equations that need to be solved to find the equilibrium polarization in the nano-box are then found from Eq. (70) with $f = f_{\text{cubic}}$. The steps will not be shown here but the cyclic operator can make the working easier and it is straightforward to show that the required Euler-Lagrange equations are given by

$$
\alpha_1 (T - T_C) P_i + P_i \sum_{j=2}^{3} \alpha_j \left(P^2 \right)^{j-1} + \beta_2 P_i \sum_{j=1}^{2} \sigma^j \left(P_j^2 \right)
$$

$$
+ \beta_3 P_i \left[2 P_i^2 \sum_{j=1}^{2} \sigma^j \left(P_i^2 \right) + \sum_{j=2}^{2} \sigma^j \left(P_j^4 \right) \right] + \delta \nabla^2 P_i + \frac{1}{\epsilon_0 \epsilon_\infty} \left(P_i - \langle P_i \rangle \right), \quad i = 1, 2, 3. \quad (73)
$$

The surface term is given by Eq. (45), but here we assume that $d(\mathbf{x}|_S)$ is constant over each surface of the box. The boundary conditions in Eq. (71) can then be shown to reduce to

$$
\mp \frac{\partial P_i}{\partial x_i} + \frac{1}{d_i^{\mp}} = 0 \ \forall \mathbf{x} \in S_{x_i = l_i^{\mp}} \quad i = 1, 2, 3. \tag{74}
$$

For a free standing box it is reasonable to have $d_i^{\mp} = d \ \forall i = 1, 2, 3$.

The solution of Eqs. (73) and (74) has not yet been obtained but progress is being made on it and will be reported later. Eq. (73) is a nonlinear partial differential equation, and as such it is unlikely that an analytical solution can be found. Instead, a numerical approach or an approximate solution using trial functions with adjustable parameters could be employed. The box geometry is likely to be amenable to a finite difference numerical solution. Dealing with a simpler symmetry such as that discussed in the introduction would make the problem less computationally intensive but still not amenable to an analytical solution, as will now be discussed.

If Eq. (72) were reduced to a simpler form by putting $\alpha_3 = \beta_2 = \beta_3 = \gamma_3 = 0$, neglecting the depolarization terms, and arranging for the polarization to be aligned along only one coordinate axis; the problem would then be something like a three-dimensional form of the thin-film case discussed in Section 2.2. It might then be tempting to try a separation of variables approach to solving the Euler-Lagrange equations with the hope that the solution would be of the form $P(x_1) P(x_2) P(x_3)$ in which each factor is a solution of the form of the thin film solution given in Section 2.2. However any attempt to do this would fail to produce an exact solution to the three-dimensional problem due to the nonlinearity of a term cubic in the polarization. However it might be useful to treat $P(x_1) P(x_2) P(x_3)$ as an approximate function and introduce into it some variable parameters that could be optimized to find the best approximation. This idea is currently being explored but as yet there are no final conclusions.

5.2. Polarization for a ferroelectric nano-strip

The case of a strip in which confinement along two of the axes can be thought of as a special case of the box in which one side, l_1 say, approaches infinity such that $\mathbf{P}(\mathbf{x}) = \mathbf{P}(x_1, x_2, x_3) \rightarrow$

$\mathbf{P}(x_2, x_3)$, and is constant with respect to x_3 since the boundaries in this direction have no influence, which also implies that the depolarization terms only involve $\langle \mathbf{P} \rangle = (\langle P_2 \rangle, \langle P_3 \rangle)$.

Now the relevant quantity is a free energy density per unit length, since for a section of the strip between $x_1 = 0$ and $x_1 = l'_1$ the free energy is

$$G = \int_0^{l'_1} dx_1 \int_A f\left(\mathbf{x}, \mathbf{P}, \mathbf{P}_{x_2}, \mathbf{P}_{x_3}, \langle \mathbf{P} \rangle\right) dA + \int_0^{l'_1} dx_1 \int_\Gamma f_{\text{surf}}(\mathbf{x}, \mathbf{P}) d\Gamma, \qquad (75)$$

leading to an energy density

$$F = \frac{G}{l'_1} = \int_A f\left(\mathbf{x}, \mathbf{P}, \mathbf{P}_{x_2}, \mathbf{P}_{x_3}, \langle \mathbf{P} \rangle\right) dA + \int_\Gamma f_{\text{surf}}(\mathbf{x}, \mathbf{P}) d\Gamma, \qquad (76)$$

where $\int_\Gamma$ is a line integral around the path Γ that traces a rectangular cross section of the strip normal to x_1, $\int_A \equiv \int_{l_2^-}^{l_2^+} \int_{l_3^-}^{l_3^+}$, and $dA = dx_2 dx_3$.

For $f = f_{\text{cubic}}$ (Eq. (72)), the Euler-Lagrange equations are similar to Eq. (73) but P_1 is constant, and so $\nabla^2 = \frac{\partial^2}{\partial x_2^2} + \frac{\partial^2}{\partial x_3^2}$, and the depolarization term for $i = 1$ reduces to zero. Also the boundary conditions are given by Eq. (74), but with $i = 2, 3$. The solution is again best approached by numerical or other approximate methods. Work on this is in progress.

The comments made in at the end of the previous section for the ferroelectric nano-box with regard to simplifying it to a three dimensional form of the problem in Section 2.2 also apply here: for the same reason it is not possible to express an exact solution of the form $P(x_2)P(x_3)$, where P has the form of the solution given in Section 2.2. Again it may be possible to utilize such a solution as an approximate function. This needs to be investigated further.

5.3. Polarization for a ferroelectric nano-film

For a film confined in the x_3 direction, a similar argument to the one given for a nano-strip yields $\mathbf{P} = \mathbf{P}(x_3)$, $\langle \mathbf{P} \rangle = \langle P_3 \rangle \hat{\mathbf{x}}_3$, and a free energy density

$$F = \frac{G}{l'_1 l'_2} = \int_{l_3^-}^{l_3^+} f\left(x_3, \mathbf{P}, \mathbf{P}_{x_3}, \langle P_3 \rangle\right) dx_3. \qquad (77)$$

With $f = f_{\text{cubic}}$ given by Eq. (72), the Euler-Lagrange equations again take the form of Eq. (73); now with P_1 and P_2 constant so that the depolarization terms reduce to zero for $i = 1, 2$, and $\nabla^2 \to \frac{\partial^2}{\partial x_3^2}$. The boundary conditions are given by Eq. (74), with $i = 3$ and $\frac{\partial}{\partial x_3} \to \frac{d}{dx_3}$. An analytical solution, even for this one-dimensional case is still not likely to exist unless f is reduced to the simpler higher symmetry form for second order transitions and in-plane polarization discussed in Section 2.2. However it is amenable to numerical solution or other approximate methods. A similar thin film problem, but involving strain in the ferroelectric as well, which we do no consider, has been solved numerically using a finite difference method by Wang and Zhang[7].

6. Conclusion

The foundations of Landau-Devonshire theory have been introduced and it has been used to develop a general formulation for the calculation of the spontaneous polarization in a ferroelectric nano-box which is confined to the nanoscale in all three dimensions, and for which the influence of the surfaces is expected to be more pronounced than for large scales. From the formalism for the box calculations it has been shown how to deal with a nano-strip which is confined to the nanoscale in two dimensions and to a nano-film with such confinement in only one dimension. The thin film case has been fairly well studied, but much less work has been done on the other geometries, box and strip. Such work is timely given the increasing use of nanoscale structures in electronic devices which include ferroelectric materials, for example ferroelectric random access memories.

A particular example of how the formalism can applied to a ferroelectric crystal of cubic symmetry has been given for all three geometries. Many ferroelectric materials have this symmetry. However the general formalism presented is not restricted to this symmetry, any symmetry can be handled through knowledge of which terms in the free energy needed to be dropped, the criterion being that only terms invariant under the symmetry operations of the symmetry group are allowed.

Future work will involve investigating the numerical or approximate function solutions to the Euler-Lagrange equations which need to be solved in order to be able to plot the spontaneous polarization. Such work is in progress and will be reported in due course.

Acknowledgements

The author would like to thank Manas Kumar Halder for useful and enlightening discussions on several of the ideas presented in this chapter.

Author details

Jeffrey F. Webb

Faculty of Engineering, Computing and Science, Swinburne University of Technology, Sarawak Campus, Kuching, Sarawak, Malaysia

References

[1] P. Chandra and P. B. Littlewood. A Landau primer for ferroelectrics. In K. Rabe, Ch. H. Ahn, and J. M. Triscone, editors, *Physics of Ferroelectrics*, volume 105 of *Topics in Applied Physics*, page 69. Springer, Heidelberg, 2007.

[2] J. F. Scott. The physics of ferroelectric ceramic thin films for memory applications. *Ferroelectr. Rev.*, 1:1, 1998.

[3] D. R. Tilley and B. Zeks. Landau theory of phase transitions in thick films. *Solid State Commun.*, 49:823, 1984.

[4] D. R. Tilley. Phase transitions in thin films. In N. Setter and E. L. Colla, editors, *Ferroelectric Ceramics*, page 163. Birkhäuser Verlag, Berlin, 1993.

[5] D. R. Tilley. Finite-size effects on phase transitions in ferroelectrics. In Carlos Paz de Araujo, J. F. Scott, and G. W. Taylor, editors, *Ferroelectric Thin Films: Synthesis and Basic Properties*, Integrated Ferroelectric Devices and Technologies, page 11. Gordon and Breach, Amsterdam, 1996.

[6] J. F. Webb. Theory of size effects in ferroelectric ceramic thin films on metal substrates. *J. Electroceram.*, 16:463, 2006.

[7] J. Wang and T. Y. Zhang. Influence of depolarization field on polarization states in epitaxial ferroelectric thin films with nonequally biaxial misfit strains. *Physical Review B*, 77:014104–1 to 014104–7, 2008.

[8] Anna N. Morozovska, Maya D. Glinchuk, and Eugene A. Eliseev. Phase transitions induced by confinement of ferroic nanoparticles. *Physical Review B*, 76:014102–1 to 014102–13, 2007.

[9] M. E. Lines and A. M. Glass. *Principles and Applications of Ferroelectrics and Related Materials*. Clarendon, Oxford, UK, 1977.

[10] B. A. Strukov and A. P. Lenanyuk. *Ferroelectric Phenomena in Crystals*. Springer, Berlin, 1998.

[11] R. C. Tolman. *The Principles of Statistical Thermodynamics*. Clarendon Press, Oxford, 1938.

[12] J. C. Toledano and P. Toledano. *The Landau Theory of Phase Transitions*. World Scientific, Singapore, 1987.

[13] R. Blinc, S. Detoni, and M. Pintar. Nature of the ferroelectric transition in triglycine sulfate. *Physical Review Letters*, 124:1036–1038, 1961.

[14] K. Itoh and T. Mitsui. Studies of the crystal structure of triglycine sulfate in connection with its ferroelectric phase transition. *Ferroelectrics*, 5:235–251, 1973.

[15] M. G. Cottam, D. R. Tilley, and B. Zeks. Theory of surface modes in ferroelectrics. *J. Phys C*, 17:1793–1823, 1984.

[16] R. Courant and D. Hilbert. *Methods of Mathematical Physics*, volume 1. Interscience, New York, 1953.

[17] X. Gerbaux and A. Hadni. *Static and dynamic properties of ferroelectric thin film memories*. PhD thesis, University of Colorado, 1990.

[18] E. K. Tan, J. Osman, and D. R. Tilley. First-order phase transitions in ferroelectric films. *Solid State Communications*, 116:61–65, 2000.

[19] K. H. Chew, L. H. Ong, J. Osman, and D. R. Tilley. Theory of far-infrared reflection and transmission by ferroelectric thin films. *J. Opt. Soc. Am B*, 18:1512, 2001.

[20] J. F. Webb. Harmonic generation in nanoscale ferroelectric films. In Mickaël Lallart, editor, *Ferroelectrics - Characterization and Modeling*, chapter 26. Intech, Rijeka, Croatia, 2011.

[21] M. Abramowitz and I. A. Stegun, editors. *Handbook of Mathematical Functions*. Dover, New York, 1972.

[22] R. Kretschmer and K. Binder. Surface effects on phase transitions in ferroelectrics and dipolar magnets. *Phys. Rev. B*, 20:1065–1076, 1979.

[23] M.D. Glinchuk, E.A. Eliseev, and V.A. Stephanovich. Ferroelectric thin film properties—depolarization field and renormalization of a "bulk" free energy coefficients. *J. Appl. Phys.*, 93:1150–1159, 2003.

[24] D. L. Mills. *Nonlinear Optics*. Springer, Berlin, second edition, 1998.

[25] E. W. Honson. On the fundamental lemma of the calculus of variations and on some related theorems. In *Proceedings of the London Mathematical Society*, volume s2-11, pages 17–28, 1913.

[26] J. Nye. *Physical Properties of Crystals. Their Representation by Tensors and Matrices*. Oxford University Press, Oxford, 1964.

Chapter 8

The Influence of Vanadium Doping on the Physical and Electrical Properties of Non-Volatile Random Access Memory Using the BTV, BLTV, and BNTV Oxide Thin Films

Kai-Huang Chen, Chien-Min Cheng, Sean Wu, Chin-Hsiung Liao and Jen-Hwan Tsai

Additional information is available at the end of the chapter

1. Introduction

Recently, the various functional thin films were widely focused on the applications in non-volatile random access memory (NvRAM), such as smart cards and portable electrical devices utilizing excellent memory characteristics, high storage capacity, long retention cycles, low electric consumption, non-volatility, and high speed readout. Additionally, the various non-volatile random access memory devices such as, ferroelectric random access memory (FeRAM), magnetron memory (MRAM), resistance random access memory (RRAM), and flash memory were widely discussed and investigated [1-9]. However, the high volatile pollution elements and high fabrication cost of the complex composition material were serious difficult problems for applications in integrated circuit semiconductor processing. For this reason, the simple binary metal oxide materials such as ZnO, Al_2O_3, TiO_2, and Ta_2O_5 were widely considered and investigated for the various functional electronic product applications in resistance random access memory devices [10-12].

The (ABO_3) pervoskite and bismuth layer structured ferroelectrics (BLSFs) were excellent candidate materials for ferroelectric random access memories (FeRAMs) such as in smart cards and portable electric devices utilizing their low electric consumption, nonvolatility, high speed readout. The ABO_3 structure materials for ferroelectric oxide exhibit high remnant polarization and low coercive filed. Such as $Pb(Zr,Ti)O_3$ (PZT), $Sr_2Bi_2Ta_2O_9$ (SBT), $SrTiO_3$ (ST), $Ba(Zr,Ti)O_3$ (BZ1T9), and $(Ba,Sr)TiO_3$ (BST) were widely studied and discussed for large storage capacity FeRAM devices. The (Ba,Sr)TiO$_3$ and Ba(Ti,Zr)O$_3$ ferroelectric materials were also expected to substitute the PZT or SBT memory materials and improve the environmental pollution because of their low pollution problem [9-15]. In addition, the high dielectric constant and low leakage current density of zirconium and strontium-doped BaTiO$_3$ thin films were applied for the further application in the high density dynamic random access memory (DRAM) [16-20].

Bismuth titanate system based materials were an important role for FeRAMs applications. The bismuth titanate system were given in a general formula of bismuth layer structure ferroelectric, $(Bi_2O_2)^{2+}(A_{n-1}B_nO_{3n+1})^{2-}$ (A=Bi, B=Ti). The high leakage current, high dielectric loss and domain pinning of bismuth titanate system based materials were caused by defects, bismuth vacancies and oxygen vacancies. These defects and oxygen vacancies were attributed from the volatilization of Bi$_2$O$_3$ of bismuth contents at elevated temperature [21-23].

1.1. ABO$_3$ pervoskite structure material system

For ABO$_3$ pervoskite structure such as, BaTiO$_3$ and BZ1T9, the excellent electrical and ferroelectric properties were obtained and found. For SOP concept, the ferroelectric BZ1T9 thin film on ITO substrate were investigated and discussed. For crystallization and grain grow of ferroelectric thin films, the crystal orientation and preferred phase of different substrates were important factors for ferroelectric thin films of MIM structures.

The XRD patterns of BZ1T9 thin films with 40% oxygen concentration on Pt/Ti/SiO$_2$/Si substrates from our previous study were shown in Fig. 1 [24-25]. The (111) and (011) peaks of the BZ1T9 thin films on Pt/Ti/SiO$_2$/Si substrates were compared with those on ITO substrates. The strongest and sharpest peak was observed along the Pt(111) crystal plane. This suggests that the BZ1T9 films grew epitaxially with the Pt(111) bottom electrode. However, the (111) peaks of BZ1T9 thin films were not observed for (400) and (440) ITO substrates. Therefore, we determined that the crystallinity and deposition rate of BZ1T9 thin films on ITO substrates differed from those in these study [24-27].

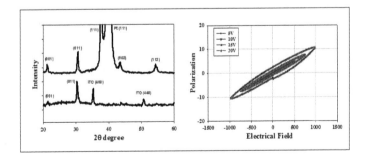

Figure 1. a) XRD patterns of as-deposited thin films on the ITO/glass and Pt substrates, and (b) p-E curves of thin films.

The polarization versus applied electrical field (*p-E*) curves of as-deposited BZ1T9 thin films were shown in Fig. 1(a). As the applied voltage increases, the remanent polarization of thin films increases from 0.5 to 2.5 $\mu C/cm^2$. In addition, the $2P_r$ and coercive field calculated and were about 5 $\mu C/cm^2$ and 250 kV/cm, respectively. According to our previous study, the BZ1T9 thin film deposited at high temperature exhibited high dielectric constant and high leakage current density because of its polycrystalline structure [24].

1.2. Bismuth Layer Ferroelectric Structure material system

The XRD patterns of as-deposited $Bi_4Ti_3O_{12}$ thin films and ferroelectric thin films under 500~700 °C rapid thermal annealing (RTA) process were compared in Fig. 2(a). From the results obtained, the (002) and (117) peaks of as-deposited $Bi_4Ti_3O_{12}$ thin film under the optimal sputtering parameters were found. The strong intensity of XRD peaks of $Bi_4Ti_3O_{12}$ thin film under the 700 °C RTA post-treatment were is found. They were (008), (006), (020) and (117) peaks, respectively. Compared the XRD patterns shown in Fig. 2, the crystalline intensity of (111) plane has no apparent increase as the as-deposited process is used and has apparent increase as the RTA-treated process was used. And a smaller full width at half maximum value (FWHM) is revealed in the RTA-treated $Bi_4Ti_3O_{12}$ thin films under the 700 °C post-treatment. This result suggests that crystal structure of $Bi_4Ti_3O_{12}$ thin films were improved in RTA-treated process.

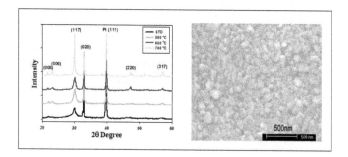

Figure 2. a) XRD patterns of as-deposited $Bi_4Ti_3O_{12}$ thin films, and (b)the SEM morphology of as-deposited $Bi_4Ti_3O_{12}$ films.

The surface morphology observations of as-deposited $Bi_4Ti_3O_{12}$ thin films under the 700 °C RTA processes were shown in Fig. 2(b). For the as-deposited $Bi_4Ti_3O_{12}$ thin films, the morphology reveals a smooth surface and the grain growth were not observed. The grain size and boundary of $Bi_4Ti_3O_{12}$ thin films increased while the annealing temperature increased to 700 °C. In RTA annealed $Bi_4Ti_3O_{12}$ thin films, the maximum grain size were about 200 nm and the average grain size is 100 nm. The thickness of annealed $Bi_4Ti_3O_{12}$ thin films were calculated and found from the SEM cross-section images. The thickness of the deposited $Bi_4Ti_3O_{12}$ thin films is about 800 nm and the deposited rate of $Bi_4Ti_3O_{12}$ thin films is about 14 nm/mim.

2. Experimental Detail

S. Y. Wu firstly reported that an MFS transistor fabricated by using bismuth titanate in 1974 [28-29]. The first ferroelectric memory device was fabricated by replacing the gate oxide of a conventional metal-oxide-semiconductor (MOS) transistor with a ferroelectric material. However, the interface and interaction problem between the silicon substrate and ferroelectric films were very important factors during the high temperature processes in 1TC structure. To overcome the interface and interaction problem, the silicon dioxide and silicon nitride films were used as the buffer layer. The low remnant polarization and high operation voltage of 1TC were also be induced by gate oxide structure with double-layer ferroelectric silicon dioxide thin films. Sugibuchi et al. provided a 50 nm silicon dioxide thin film between the $Bi_4Ti_3O_{12}$ layer and the silicon substrate [30].

The ferroelectric ceramic target prepared, the raw materials were mixed and fabricated by solid state reaction method. After mixing and ball-milling, the mixture was dried, grounded, and calcined for some time. Then, the pressed ferroelectric ceramic target with a diameter of two inches was sintered in ambient air. The base pressure of the deposited chamber was brought down 1×10^{-7} mTorr prior to deposition. The target was placed away from the $Pt/Ti/SiO_2/Si$ and SiO_2/Si substrate. For metal-ferroelectric-metal (MFM) capacitor structure, the Pt and the Ti were deposited by dc sputtering using pure argon plasma as bottom electrodes. The SiO_2 thin films were prepared by dry oxidation technology. The metal-ferroelectric-insulator-semiconductor (MFIS) and metal-ferroelectric-metal (MFM) structures were shown in Fig. 3.

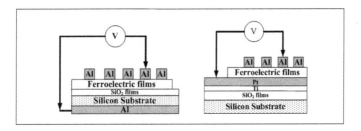

Figure 3. a) Metal-ferroelectric-insulator-semiconductor (MFIS) structure, and (b) Metal ferroelectric-metal (MFM) structure.

For the physical properties of ferroelectric thin films obtained, the thickness and surface morphology of ferroelectric thin films were observed by field effect scanning electron microscopy (FeSEM). The crystal structure of ferroelectric thin films were characterized by an X-ray diffraction (XRD) measurement using a Ni-filtered CuKα radiation. The capacitance-voltage (C-V) properties were measured as a function of applied voltage by using a Hewlett-Packard (HP 4284A) impedance gain phase analyzer. The current curves versus the applied voltage (I-V characteristics) of the ferroelectric thin films were measured by a Hewlett-Packard (HP 4156) semiconductor parameter analyzer.

176	Handbook of Ferroelectrics

3. Results and Discussion

3.1. Large memory window in the vanadium doped $Bi_4Ti_3O_{12}$ (BTV) thin films

The XRD pattern was used to identify the crystalline structures of as-deposited BTV thin films, labeled "vanadium doped at 550 °C," with various depositing parameters. From the XRD pattern, we found that the optimal deposition parameters of as-deposited BTV thin films were RF power of 130 W, chamber pressures of 10 mtorr and oxygen concentrations of 25%. The crystalline orientations of (117), (008) and (200) planes were apparently observed in the films. It was found that all of the films consisted of a single phase of a bismuth layered structure showing the preferred (008) and (117) orientation. Both films were well c-axis oriented, but BTV thin film was more c-axis oriented than BIT, labeled "undoped at 550 °C". For the polycrystalline BTV thin films, the (117) peak was the strongest peak and the intensity of the (008) peak was 10% of (117) peak intensity. An obvious change in the orientation due to the substitution was observed except for the degree of the (117) orientation for BTV films. In addition, the XRD patterns of the as-deposited BTV thin films deposited using optimal parameters at room and 550 °C substrate temperatures were observed in Fig. 2. This result indicated that the crystalline characteristics of BTV thin films deposited at 550 °C were better than those of BTV thin films at room temperature. The crystalline and dielectric characteristics of as-deposited BTV thin films were influenced by substrate temperatures. The electrical characteristics of as-deposited BTV thin films at substrate temperatures of 550 °C under optimal parameters will be further developed.

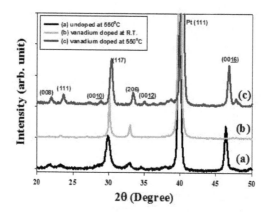

Figure 4. The XRD patterns of (a) undoped at 550 °C (b) vanadium doped at R.T. and (c) vanadium doped at 550 °C thin films deposited using optimal parameters.

In Fig. 5, circular-like grains with 150 nm width were observed with scanning electron microscopy (SEM) for as-deposited BTV thin films. From the cross-sectional SEM image, film thicknesses were measured to be 742 nm. As the depositing time increases from 30 and 60, to 120 min, the thickness of as-deposited BTV thin films increases linearly from 197 and 386, to 742 nm, respectively, as the depositing rate decreases from 6.57 and 6.43, to 6.18 nm.

Figure 5. The surface micro structure morphology of the BTV thin films deposited using optimal parameters.

Figure 6(a) compares the change in the capacitance versus the applied voltage (*C-V*) for the un-doped and vanadium doped thin films. Based on Fig. 4, the capacitances of the BIT thin films appear to increase due to the vanadium dopant. We found that the capacitances of BTV thin films increased from 1.3 to 4.5 nF. As suggested by Fig. 4, the improvement in the dielectric constants of the BTV thin films can be attributed to the compensation of the oxygen vacancy and the improvement in the B-site substitution of the ABO_3 phase in the BTV thin films [31-34].

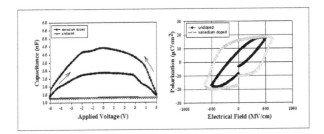

Figure 6. a) The P-E characteristics of vanadium doped and undoped thin films, and (b) The normalization C-V curves of vanadium doped and undoped thin films.

Figure 6 shows ferroelectric hysteresis loops of BIT and as-deposited BTV thin film capacitors measured with a ferroelectric tester (Radiant Technologies RT66A). The as-deposited BTV thin films, labeled "vanadium doped," clearly show ferroelectricity. The remanent po-

larization and coercive field were 23 μC/cm² and 450 kV/cm. Comparing the vanadium doped and undoped BIT thin films, the remanent polarization (2Pr) would be increased form
16μC/cm² for undoped BIT thin films to 23 μC/cm² for vanadium doped. However, the coercive field of as-deposited BTV thin films would be increased to 450 kV/cm. These results indicated that the substitution of vanadium was effective for the appearance of ferroelectricity
at 550 °C. The 2Pr value and the Ec value were larger than those reported in Refs. [35-36],
and the 2Pr value was smaller and the Ec value was larger than those reported in [37]. Based
on above results, it was found that the simultaneous substitutions for B-site are effective to
derive enough ferroelectricity by accelerating the domain nucleation and pinning relaxation
caused by B-site substitution [31-37].

The leakage current density versus applied voltage curves of as-deposited BTV thin films for
different depositing time on the MFIS structure were be found. We found that the leakage
current density of undoped BIT thin films, labeled "undoped," were larger than those of vanadium-doped BIT thin films. This result indicated that the substitution of B-site in ABO_3
perovskite structure for BTV thin films was effective in lowering leakage current density.
Besides, the thickness of BTV thin films has an apparent influence on the leakage current
density of BTV thin films, and that will have an apparent influence on the other electrical
characteristics of BTV thin films. At an electric field of 0.5 MV/cm, the leakage current density critically decreases from the 3.0 ×10⁻⁷ A/cm2 for 30 min-deposited BTV thin films to
around 3×10⁻⁸A/cm² and 2×10⁻⁸ A/cm² for 60 and 120 min-deposited BTV thin films.

Figure 7(a) show the capacitance versus applied voltage (C-V) curves of as-deposited vanadium doped BTV and un-doped BIT thin films. The applied voltages, which are first changed
from -20 to 20 V and then returned to -20 V, are used to measure the capacitance voltage characteristics (C-V) of the MFIS structures. For the vanadium doped thin films, the memory window of MFIS structure increased from 5 to 15 V, and the threshold voltage decreased
from 7 to 3 V. This result demonstrated that the lower threshold voltage and decreased oxygen vacancy in undoped BIT thin films had been improved from the C-V curves measured.

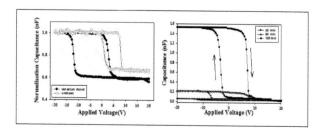

Figure 7. a) The P-E characteristics of vanadium doped and undoped thin films, and (b) The normalization C-V curves
of vanadium doped and undoped thin films.

Figure 7(b) shows the C-V curves of 30 min-deposited BTV thin films, and Figure 7(b) compares the C-V curves of BTV thin films deposited at different depositing time. Figure 7(b)
also shows that the capacitance at the applied voltage value of 0 V critically increases as the

depositing time increases. The memory window would be decreased from 15.1 and 13.4, to 10.6 V as the depositing time increased. Two reasons may be the cause that the increases capacitance of as-deposited BTV thin films in the MFIS structure for the different depositing time. First, the decrease leakage current density was attributed to the increase thickness of as-deposited BTV thin films in Fig. 4. Second, the different thickness and high dielectric constant of as-deposited BTV thin films are also an important factor. In Fig. 7, the capacitance of as-deposited BTV thin films for MFIS structure could be calculated from Eq. 1 and Eq. 2:

$$C = \varepsilon_0 \varepsilon_r \frac{A}{d} \tag{1}$$

$$\varepsilon_r = \frac{(\varepsilon_1 d_2 + \varepsilon_2 d_1)}{(d_1 + d_2)} \tag{2}$$

The $\varepsilon 1$ and d1 are the effective dielectric constant and the total thickness of silicon and SiO_2 layer. The $\varepsilon 2$ and d2 are the effective dielectric constant and the thickness of as-deposited BTV thin film layer. The relative dielectric constants of Si and SiO_2 are 11.7 and 3.9. The εr value ($\varepsilon 1$) of silicon and SiO_2 layer is much smaller than that ($\varepsilon 2$) of as-deposited BTV thin films. The $\varepsilon 1 \times d2$ is the unchanged value and the $\varepsilon 2 \times d1$ value increases with the increase of depositing time. In Eq. (2), the $\varepsilon 1 \times d2$ increases more quickly as d2 increases. In Eq. (1), the C value will increase as the εr value increases.

For memory window characteristics at applied voltage of 0 volts, the upper and lower capacitance values of as-deposited BTV thin films for 30 min depositing time were 0.056 and 0.033 nF, respectively. For 60 min and 120 min depositing time, they were 0.215~0.048 nF and 1.515 ~0.105 nF, respectively. The change ratios at zero voltage were defined in Eq.(3) from these experimental results:

$$ratio = \frac{(C_u - C_l)}{C_u} \tag{3}$$

where Cu and Cl are the upper and lower capacitance values.

The capacitance change ratios of as-deposited BTV thin films for different depositing time were 41, 73 and 93%, respectively. From above statements, the good switching characteristics of ferroelectric polarization could be attributed to memory windows ratio and the thinner thickness of as-deposited BTV thin film for the depositing time of 30 min. These results indicted the upper and lower capacitance of memory window would be decreased by lowering the thickness of SiO_2 layer.

3.2. The Influence of Lanthanum Doping on the Physical and Electrical Properties of BTV (BLTV) Ferroelectric Thin Films

For MFM structures, the crystal orientation and preferred phase of ferroelectric thin films on Pt/Ti/SiO$_2$/Si substrates was important factor. The x-ray diffraction (XRD) patterns of BLTV and BTV thin films prepared by rf magnetron sputtering were be found. From the XRD pattern, the BLTV and BTV thin film were polycrystalline structure. The (004), (006), (008), and (117) peaks were observed in the XRD pattern. All of thin films consisted of a single phase of a bismuth layered structure showing the preferred (117) orientation. All of thin films were exhibited well c axis orientation. The change in the orientation of BLTV thin films due to the substitution was not observed. The degree of the (117) orientation relative to the (001) orientation of BLTV thin films dominant was shown.

Figure 8. The surface micro structure of as-deposited (a) BTV and (b) BLTV thin films.

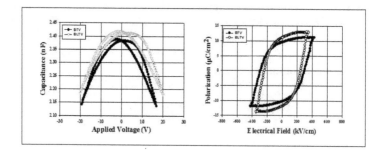

Figure 9. a)The C-V characteristics of as-deposited BTV and BLTV thin films. (b) The polarization versus electrical field characteristics of as-deposited BTV and BLTV thin films.

In Fig. 8a, rod-like and circular-board grains with 250 nm length and 150 nm width were observed with scanning electron microscopy (SEM) for as-deposited BTV films. The small grain was gold element in preparation for the SEM sample. However, the BLTV thin films exhibited a great quantity of 400 nm length and 100 nm width rod-like grain structure in Fig 8 b. The rod-like grain size of BLTV thin films was larger than those of BTV. We induced that the bismuth vacancies of BTV thin films compensate for lanthanum addition and microstructure were improved in BLTV thin films. From the cross-sectional SEM image, average thin film thicknesses for MFIS structure were about 610 nm. The average thickness of thin films for MFM structure was about 672 nm.

Figure 9(a) shows the change in the capacitance versus the applied voltage (C-V) of the BTV and BLTV thin films in MFM structure measured at 100 kHz. The applied voltages, which were first changed from -20 to 20 V and then returned to -20 V, were used to measure the capacitance voltage characteristics (CV). The BLTV thin films exhibited high capacitance than those of BTV thin films. We found that the capacitances of the lanthanum-doped BTV thin films were increased from 2.38 to 2.42 nF.

Figure 9(b) shows the p-E curves of the different ferroelectric thin films under applied voltage of 18V from the Sawyer–Tower circuits. The remanent polarization of non-doped, vanadium-doped, and lanthanum-doped ferroelectric thin films linearly was increased from 5, 10 to 11 $\mu C/cm^2$, respectively. The coercive filed of non-doped, vanadium-doped, and lanthanum-doped ferroelectric thin films were about 300, 300, and 250 kV/cm, respectively. The ferroelectric properties of lanthanum-doped and vanadium-doped BIT thin films were improved and found.

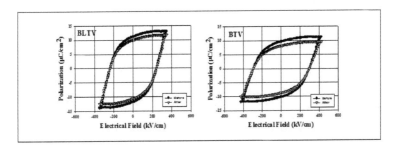

Figure 10. a)The C-V characteristics of as-deposited BTV and BLTV thin films. (b) The polarization versus electrical field characteristics of as-deposited BTV and BLTV thin films.

The fatigue characteristics for ferroelectric thin films were the time dependent change of the polarization state. After a long time, the polarization loss of ferroelectric thin films was affected by oxygen vacancies, defect, and space charge in the memory device. Figure 10 shows the polarization versus electrical field (p-E) properties of the different ferroelectric thin films before and after the switching of 10^9 cycles. The remnant polarization loss of BLTV and BTV thin films were about 9% and 15% of initial polarization value, respectively. The remnant polarization of BLTV thin films were little changed after the switching cycles. The fatigue behavior and domain pinning were improved by lanthanum and vanadium addition in BIT thin films. To reduce bismuth defect and oxygen vacancy, the high-valence cation substituted for the A-site of BLTV thin films were observed.

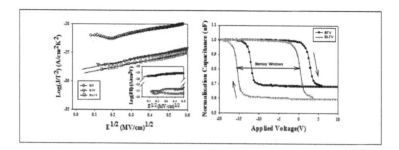

Figure 11. a) The (J/T²) and (J/E) versus E^{1/2} curves of as-deposited BIT, BTV, and BLTV thin films. (b) The normalization capacitance versus applied voltage curves of as-deposited BTV and BLTV thin films.

From the J-E curves, the conduction mechanism of as-deposited thin films was also proved the bismuth defect and oxygen vacancy results. For MFIS structure, the interface and interaction problem between of the silicon and ferroelectric films was serious important. The effect of the bound polarization charge from the carrier in the silicon substrate was observed. To overcome this problem, the insulator films of buffer layer were used for MFIS structure [28-30, 37-38].

Figure 11(a) shows the leakage current density versus electrical field (J-E) characteristics of as-deposited BTV and BLTV thin films for MFIS structure. The leakage current density of the as-deposited BLTV thin films were about one order of magnitude lower than those of the non-doped lanthanum thin films. However, the lanthanum and vanadium doped BIT thin films were lower than those of BIT thin films. We suggested that low leakage current density attributed to substituting a bismuth ion with a lanthanum ion at A-site for lanthanum doped BTV thin films. To discuss bismuth ion substituting by lanthanum ion effect, the leakage current versus electrical field curves of BLTV thin films were fitted to Schottky emission and Poole–Frankel transport models. The inset of Fig. 11(a) shows the JE characteristics for BIT, BTV, and BLTV thin films in terms of J/E as vertical axis and E^{1/2} as horizontal axis. The fitting curves were straight in this figure and a JE curve of thin films was the Poole–Frankel emission model. The high leakage current was attributed to the bismuth vacancies and oxygen vacancies of as-deposited BIT thin films [39–42]. Park et al. also suggested that the enhanced stability of TiO_6 octahedra against oxygen vacancies for fatigue resistance of lanthanum doped BIT thin films [31].

In a previous study, the low threshold voltage of ferroelectric thin films was attributed by bismuth and oxygen vacancy [9]. The threshold voltage for the lanthanum-doped BTV thin films of MFIS structure was improved from 5 to 3 V. The memory functional effect and depletion delay of the MFIS structure was caused by remanent polarization of ferroelectric thin films in CV curves. In this study, the memory window was increased from 15 to 18 V. The large memory window of lanthanum-doped BTV thin films was also proved by p-E curves in Fig. 11(b). As a result, the improvement in the capacitance of the BLTV thin films was attributed to the compensation of the oxygen vacancy of the BLSF structure. Additionally, the memory window of ferroelectric thin films was changed by the sweeping speed. The ferro-electric capacitance and threshold voltage of MFIS structure slightly decreases as the sweeping speed increases. That was influenced by the mobile ions and charge injection between as-deposited ferroelectric thin films and metal electrode as the sweeping speed increased [9].

3.3. The Influence of Neodymium Doping on the Physical and Electrical Properties of BTV (BNTV) Ferroelectric Thin Films

Figure 12(a) shows x-ray diffraction patterns of the as-deposited ferroelectric thin films for different oxygen concentration on ITO substrate. From the XRD patterns, we found that the ferroelectric thin films exhibited polycrystalline structure. In addition, the (117), (008), and (220) peaks were observed in the XRD pattern. The intensity of the (117) peak of the ferro-electric thin films increases linearly as the oxygen concentration increases from 0 to 40%. The intensity of the (117) peak of the as-deposited ferroelectric thin films decreases at oxy-gen concentration from 40 to 60%. As shown in Fig. 3, the (117) preferred phase and smallest full-width-half-magnitude (FWHM) value were exhibited by the as-deposited ferroelectric thin film with the 40% oxygen concentration. The polycrystalline structure of the as-deposit-ed $(Bi_{3.25}Nd_{0.75})(Ti_{2.9}V_{0.1})O_{12}$ ferroelectric thin film was optimal at 40% oxygen concentration.

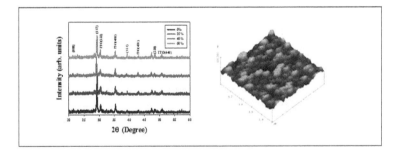

Figure 12. a) The x-ray diffraction patterns of the as-deposited $(Bi_{3.25}Nd_{0.75})(Ti_{2.9}V_{0.1})O_{12}$ ferroelectric thin film for the different oxygen concentration. (b) The surface morphology of the as-deposited $(Bi_{3.25}Nd_{0.75})(Ti_{2.9}V_{0.1})O_{12}$ ferroelectric thin film for 40% oxygen concentration.

Besides, the interface between an electrode and the as-deposited $(Bi_{3.25}Nd_{0.75})(Ti_{2.9}V_{0.1})O_{12}$ thin films was an important factor that seriously influences the physical and electrical properties of the MIM capacitor structure. Therefore, the surface roughness of the as-deposited $(Bi_{3.25}Nd_{0.75})(Ti_{2.9}V_{0.1})O_{12}$ thin films for 40% oxygen concentration was absolutely determined and calculated. Figure 12(b) shows the surface roughness of the as-deposited $(Bi_{3.25}Nd_{0.75})$ $(Ti_{2.9}V_{0.1})O_{12}$ ferroelectric thin film from the AFM images. The roughness of the ferroelectric thin film were 4.278nm. The surface roughness of the as-deposited ferroelectric thin films in-creases with the oxygen concentration. Therefore, we assume that the surface roughness of the as-deposited ferroelectric thin films increases due to an increase in the crystallinity with oxygen concentration.

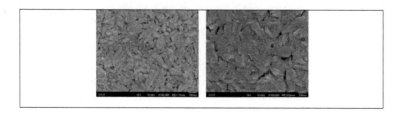

Figure 13. The surface morphology of the as-deposited $(Bi_{3.25}Nd_{0.75})(Ti_{2.9}V_{0.1})O_{12}$ ferroelectric thin film for 40% oxygen concentration. The surface micro-structure of the as-deposited $(Bi_{3.25}Nd_{0.75})(Ti_{2.9}V_{0.1})O_{12}$ ferroelectric thin film for (a) 25% and (b) 40% oxygen concentration.

From the SEM images in Fig. 13, the surface morphology and grain size of the as-deposited $(Bi_{3.25}Nd_{0.75})(Ti_{2.9}V_{0.1})O_{12}$ thin films for 25 and 40% oxygen concentration were observed. The grain size of the as-deposited $(Bi_{3.25}Nd_{0.75})(Ti_{2.9}V_{0.1})O_{12}$ thin films were about 110 nm and 50 nm, respectively. We deduced that grain size changed caused by the different oxygen xoncentration.

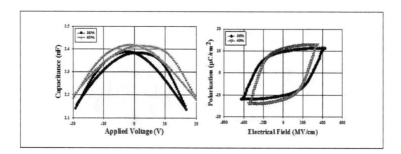

Figure 14. a) The capacitance versus applied voltage (C-V) properties of the as-deposited $(Bi_{3.25}Nd_{0.75})(Ti_{2.9}V_{0.1})O_{12}$ ferroelectric thin film. (b) The polarization versus applied electrical field (p-E) properties of the as-deposited $(Bi_{3.25}Nd_{0.75})(Ti_{2.9}V_{0.1})O_{12}$ ferroelectric thin film.

Figure 14(a) shows the C-V characteristics measured with the MFM capacitor structure for the as-deposited $(Bi_{3.25}Nd_{0.75})(Ti_{2.9}V_{0.1})O_{12}$ thin films deposited under various oxygen concentrations. The applied bias voltage is adjusted from -20 to 20V. The capacitance of ferroelectric thin films first increases with the increase of oxygen concentration and reaches the maximum value in the 40 % oxygen atmosphere. Then the capacitance apparently decreases in the further increase of oxygen to 60 %. This variation of capacitance has the similar results with the XRD patterns and the AFM images. The polarization versus applied electrical field (p-E) curves of the as-deposited $(Bi_{3.25}Nd_{0.75})(Ti_{2.9}V_{0.1})O_{12}$ ferroelectric thin film at the frequency of 1kHz were shown in fig. 14(b). From the p-E curves results, the remnant polarization of ferroelectric thin films were 11 $\mu C/cm^2$, as the coercive filed of 220 kV/cm. In addition, the remnant polarization and coercive filed of ferroelectric thin films for 40% oxygen concentration were about 10 $\mu C/cm^2$ and 300 kV/cm. From the experimental measurement, this result was attributed to the suitable oxygen concentration sample as compared to that of the as-deposited ferroelectric thin film. As the oxygen/argon mixtures were used as the depositing atmosphere, the defects and oxygen vacancies in ferroelectric thin films were filled and compensated by oxygen gas, and the leakage current density were decreased. The low leakage current density will reveal in the 40%-oxygen-deposited ferroelectric thin films. For that the capacitance will be increased and the leakage current density will be decreased. As the applied voltage of 15V was used, the leakage current density of ferroelectric thin films deposited at 40% oxygen concentration is about $1\times10^{-9}A/cm^2$.

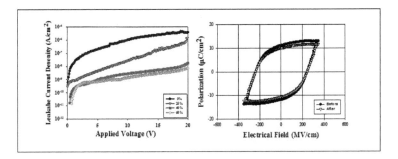

Figure 15. a) The leakage current density versus applied voltage (J-E) properties of the as-deposited $(Bi_{3.25}Nd_{0.75})$ $(Ti_{2.9}V_{0.1})O_{12}$ ferroelectric thin film.(b) The retention and fatigue properties of the as-deposited ferroelectric thin film corresponding hysteresis loop before and after fatigue test.

The retention and fatigue properties for the as-deposited $(Bi_{3.25}Nd_{0.75})(Ti_{2.9}V_{0.1})O_{12}$ ferroelectric thin films were the time dependent change of the polarization state. After long time testing, the polarization loss of the as-deposited ferroelectric thin films was affected by oxygen vacancies, defect, and space charges in the memory device test. Figure 15 shows the polarization versus electrical field (p-E) properties of the as-deposited $(Bi_{3.25}Nd_{0.75})(Ti_{2.9}V_{0.1})O_{12}$ ferroelectric thin films before and after the switching of 10^9 cycles. The remnant polarization loss of ferroelectric thin films was about 9% of initial polarization value, respectively. The remnant polarization of ferroelectric thin films was little changed after the test cycles. The fatigue behavior and domain pinning were improved by neodymium and vanadium addition in BIT thin films. To improve bismuth and oxygen vacancy, the high-valence cation substituted for the A-site of $(Bi_{3.25}Nd_{0.75})(Ti_{2.9}V_{0.1})O_{12}$ thin films were observed.

3.4. Bipolar Resistive Switching Properties of Transparent Vanadium Oxide (V_2O_5) Resistive Random Access Memory

Figure 16(b) shows x-ray diffraction patterns of the as-deposited vanadium oxide thin films for 60% oxygen concentration on ITO substrate prepared by different sintering temperature. From the XRD patterns, we found that the vanadium oxide thin films exhibited polycrystalline structure. In addition, the (110), (222), and (400) peaks were observed in the XRD pattern. The intensity of the (110) peak of the thin films increases linearly as the sintering temperature increases from 400 to 550 °C. The intensity of the (110) peak of the as-deposited thin films decreases at sintering temperature from 550 to 600 °C. As shown in Fig. 16, the (110) preferred phase and smallest full-width-half-magnitude (FWHM) value were exhibited by the as-deposited vanadium oxide thin film with the sintering temperature of 550 °C. The polycrystalline structure of the as-deposited vanadium oxide thin film was optimal at 550 °C sintering temperature.

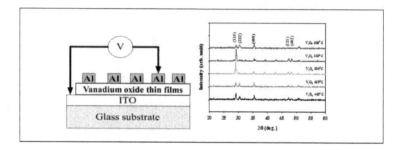

Figure 16. a) Metal-insulator-metal (MIM) structure using as-deposited vanadium oxide thin films. (b) The x-ray diffraction patterns of the as-deposited vanadium oxide thin films on ITO substrate for different sintering temperature.

The thickness of as-depoisted vandium oxide thin films for different sintering temperature was determined by SEM morphology. As the oxygen concentration increases from 0 to 60%, the thickness of as-deposited vandium oxide thin films linearly decreases. In addition, the deposition rate of as-deposited vanadium oxide thin films with 60% oxygen concentration was 2.62 nm/min. The decreases in the deposition rate and thickness of as-depoisted vanadium oxide thin films might be affected by the decrease in Ar/O_2 ratio. The Ar/O_2 ratio was adjusted using argon gas to generate the plasma on the surface of the as-deposited vanadium oxide ceramic target during sputtering. Figure 17 shows the surface morphology for the as-deposited and 500 °C sintered vanadium oxide thin films. We found that the grain size of 500 °C sintered vanadium oxide thin films were larger than others. The better resistance properties might be caused by this reason.

Figure 18 shows the current vsersus applied voltage (*I-V*) properties of vanadium oxide thin films for the different sintering temperature. After the staring forming process, the device reached a low resistance state (LRS) and high resistance state (HRS). By sweeping the bias to negative over the reset voltage, a gradual decrease of current was presented to switch the cells from LRS to HRS (reset process). Additionally, the cell turns back to LRS while applying a larger positive bias than the set voltage (set process). All of the vanadium oxide thin film were exhibit the bipolar behavior. The *I-V* properties of as-deposited vanadium oxide thin films of 60% oxygen concentration was about 1×10^{-4}A/cm^2 when an applied electrical voltage of 0.1V. During the rf sputtering deposition process, oxygen vacancies appear in the as-deposited vanadium oxide thin films. The defects and oxygen vacancies of as-deposited vanadium oxide thin films were filled and compensated for to different extents at different oxygen concentrations. In addition, the smallest leakage current density of as-deposited vanadium oxide thin films was obtained at an oxygen concentration of 40%. As shown in Fig. 18, the high leakage current density and thin films of as-deposited vanadium oxide thin films for 60% oxygen concentration were attributed to low argon sputtering gas concentration.

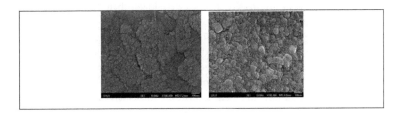

Figure 17. The surface morphology for (a) as-deposited (b) 500°C sintered vanadium oxide thin films.

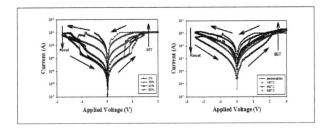

Figure 18. a) Typical *I-V* characteristics of vanadium oxide thin films for the different oxygen concentration. (b) Typical *I-V* characteristics of the as-deposited vanadium oxide thin films for different sintering temperature.

In addition, The transport current of the vanadium oxide thin films decreases linearly as the sintering temperature increases from 450 to 500 °C. The transport current of the as-deposited vanadium oxide thin films increases at sintering temperature from 500 to 550 °C. We found the as-deposited vanadium oxide thin films prepared by 500 °C sintering temperature were exhibited the large the on/off ratio resistance properties. In addition, the switching cycling was measured another type of reliability and retention characteristics were observed. There was a slight flucation of resistance in the HRS and LRS states, and the stable bipolar witching property was observed during 20 cycles. The results show remarkable reliability performance of the resistance random access memory devices for nonvolatile memory applications.

4. Conclusion

In conclusion, BIT, BTV, BNTV, and BLTV thin films were prepared by rf magnetron sputtering. We confirmed that all thin films on Pt/Ti/SiO$_2$/Si substrate well crystallized by XRD analysis. The BNTV, BLTV, and BTV shows clear ferroelectricity form the p-E curves. The remnant polarization properties of BLTV thin film decreased by 9%, while that of the BTV decreased by 15% after the fatigue test with 10^9 switching cycles. Fatigue behavior in ferroelectric capacitors was attributed to oxygen and bismuth vacancies. The leakage current density of as-deposited BLTV and BTV thin films were lower than those of BIT, which were were attributed to the decrease of oxygen and ismuth vacancies after vanadium and lanthanum addition. The conduction mechanism of as-deposited BTV and BLTV thin films were also proved these results in J-E curves. We indicated that small ions substitution for A and B site of BLSF structure was effective decreased the oxygen and bismuth vacancies. Finally, the low threshold voltage and memory window of BLTV thin films were improved from the C-V curves measured. In addition, the metal oxide thin films such as, TiO_2, Ta_2O_5, Al_2O_3, and CuO were widely investigated and discussed for applications in nonvolatile resistive random access memory (RRAM) devices. The nonvolatile resistive random access memory (RRAM) devices were well developed and studied because of their structural simplicity, high density and low power, read / write speed (about $10^1 \sim 10^3$ns), high operating cycles (> 10^{13}) and other non-volatile advantages. Therefore, the electrical switching properties of the vanadium oxide thin films for nonvolatile resistive random access memory (RRAM) device were observed.

Acknowledgements

The authors will acknowledge to Prof. Ting-Chang Chang and Prof. Cheng-Fu Yang. Additionally, this work will acknowledge the financial support of the National Science Council of the Republic of China (NSC 99-2221-E-272-003) and (NSC 100-2221-E-272-002).

Author details

Kai-Huang Chen[1], Chien-Min Cheng[2], Sean Wu[3], Chin-Hsiung Liao[4] and Jen-Hwan Tsai[4]

1 Department of Electronics Engineering/Tung-Fang Design University/Taiwan,R.O.C., Taiwan

2 Department of Electronic Engineering/Southern Taiwan University of Science and Technology/Taiwan, R.O.C., Taiwan

3 Department of Electronics Engineering/Tung-Fang Design University/Taiwan,R.O.C., Taiwan

4 Department of Mathematics and Physics/Chinese Air Force Academy/Taiwan, R.O.C., Taiwan

References

[1] Shannigrahi, R., & Jang, H. M. (2001). *Appl. Phys. Lett.*, 79, 1051.

[2] Hong, S. K., Suh, C. W., Lee, C. G., Lee, S. W., Hang, E. Y., & Kang, N. S. (2000). *Appl. Phys. Lett.*, 77-76.

[3] Xiong, S. B., & Sakai, S. (1999). *Appl. Phys. Lett.*, 75.

[4] Kim, J. S., & Yoon, S. G. (2000). *Vac. Soc. Technol.*, B, 18, 216.

[5] Wu, T. B., Wu, C. M., & Chen, M. L. (1996). *Appl. Phys. Lett.*, 69, 2659.

[6] Chen, K. H., Chen, Y. C., Yang, C. F., & Chang, T. C. (2008). *J. Phys. Chem. Solids*, 69, 461.

[7] Yang, C. F., Chen, K. H., Chen, Y. C., & Chang, T. C. (2007). *IEEE Trans. Ultrason. Ferroelectr. Freq. Control*, 54.

[8] Yang, C. F., Chen, K. H., Chen, Y. C., & Chang, T. C. (2008). *Appl. Phys. A-Mater. Sci. Process*, 90.

[9] Chen, K. H., Chen, Y. C., Chen, Z. S., Yang, C. F., & Chang, T. C. (2007). *Appl. Phys. A-Mater. Sci. Process*, 89, 533.

[10] Wu, S., Lin, Z. X., Lee, M. S., & Ro, R. (2007). *Appl. Phys. Lett.*, 102, 084908.

[11] Lee, S., Song, E. B., Kim, S., Seo, D. H., Seo, S., Won, K. T., & Wang, K. L. (2012). *Appl. Phys. Lett.*, 100, 023109.

[12] Yang, Y., Jin, L., & Yang, X. D. (2012). *Appl. Phys. Lett.*, 100, 031103.

[13] Scotta, J. F., Paz, C. A., & de Araujoa, B. M. (1994). *J. Alloys, Compounds*, 211 451.

[14] Araujo, C. A., Cuchiaro, J. D., Mc Millian, L. D., Scott, M. C., & Scott, J. F. (1995). *Nature (London)*, 374, 627.

[15] Park, B. H., Kang, B. S., Bu, S. D., Noh, T. W., Lee, J., & Jo, W. (1999). *Nature (London)*, 401, 682.

[16] Leu, C. C., Yao, L. R., Hsu, C. P., & Hu, C. T. (2010). *J. Electrochem. Soc.*, 157, 3, G85.

[17] Chen, K. H., Chen, Y. C., Yang, C. F., & Chang, T. C. (2008). *J. Phys. Chem. Solids*, 69.

[18] Yang, C. F., Chen, K. H., Chen, Y. C., & Chang, T. C. (2007). *IEEE Trans. Ultrason. Ferroelectr. Freq. Control*, 54 1726.

[19] Yang, C. F., Chen, K. H., Chen, Y. C., & Chang, T. C. (2008). *Appl. Phys. A*, 90 329.

[20] Chen, K. H., Chen, Y. C., Chen, Z. S., Yang, C. F., & Chang, T. C. (2007). *Appl. Phys. A*, 89 533.

[21] Watanabe, T., Funakubo, H., Osada, M., Noguchi, Y., & Miyayama, M. (2002). *Appl. Phys. Lett.*, 80 1.

[22] Kim, S. S., Song, T. K., Kim, J. K., & Kim, J. (2002). *J. Appl. Phys.*, 92 4.

[23] Noguchi, Y., & Miyayama, M. (2001). *Appl. Phys. Lett.*, 78 13.

[24] Velu, G., Legrand, C., Tharaud, O., Chapoton, A., Remiens, D., & Horowitz, G. (2001). *Appl. Phys. Lett.*, 79 659.

[25] Chen, K. H., Yang, C. F., Chang, C. H., & Lin, Y. J. (2009). *Jpn. J. Appl. Phys.*, 48 091401.

[26] Miao, J., Yuan, J., Wu, H., Yang, S. B., Xu, B., Cao, L. X., & Zhao, B. R. (2001). *Appl. Phys. Lett.*, 90 022903.

[27] Chen, K. H., Chen, Y. C., Chia, W. K., Chen, Z. S., Yang, C. F., & Chung, H. H. (2008). *Key Eng. Mater.*, 368-372.

[28] Wu, S. Y. (1974). *IEEE Trans. Electron Devices*, 21 499.

[29] Wu, S. Y. (1976). *Ferroelectrics*, 11 379.

[30] Sugibuchi, K., Kurogi, Y., & Endo, N. (1975). J. Appl. Phys. 46 2877. .

[31] Park, B. H., Kang, B. S., Bu, S. D., Noh, T. W., Lee, L., & Joe, W. (1999). *Nature (London)*, 401 682.

[32] Noguchi, Y., Miwa, I., Goshima, Y., & Miyayama, M. (2000). *Jpn. J. Appl. Phys*, 39 L1259.

[33] Noguchi, Y., & Miyayama, M. (2001). *Appl. Phys. Lett.*, 78 1903.

[34] Friessnegg, T., Aggarwal, S., Ramesh, R., Nielsen, B., Poindexter, E. H., & Keeble, D. J. (2000). *Appl. Phys. Lett.*, 77 127.

[35] Watanabe, T., Funakubo, H., Osada, M., Noguchi, Y., & Miyayama, M. (2002). *Appl. Phys. Lett.*, 80 1.

[36] Kim, S. S., Song, T. K., Kim, J. K., & Kim, J. (2002). *J. Appl. Phys.*, 92, 4.

[37] Noguchi, Y., & Miyayama, M. (2001). *Appl. Phys. Lett.*, 78 13.

[38] Rost, T. A., Lin, H., & Rabson, T. A. (1991). *Appl. Phys. Lett.*, 59 3654.

[39] Rost, T. A., Lin, H., Rabson, T. A., Baumann, R. C., & Callahan, D. C. (1991). IEEE Trans. Ultrason. Ferroelectr. Freq. Control, , 38

[40] Fleischer, S., Lai, P. T., & Cheng, Y. C. (1994). *J. Appl. Phys.*, 73 8353.

[41] Mihara, T., & Watanabe, H. (1995). *Part I, Jpn. J. Appl. Phys.*, 34 5664.

[42] Lin, Y. B., & Lee, J. Y. (2000). *J. Appl. Phys.*, 87 1841.

Chapter 9

Emerging Applications of Ferroelectric Nanoparticles in Materials Technologies, Biology and Medicine

Yuriy Garbovskiy, Olena Zribi and
Anatoliy Glushchenko

Additional information is available at the end of the chapter

1. Introduction

Consider an insulating system with non-zero spontaneous polarization P_s (dielectric dipole moment per unit volume). If an applied external electric field E that is greater than the so-called coercive field E_c can reverse P_s then our system is a ferroelectric system. Ferroelectricity has a long and exciting history described in [1,2]. In the beginning of its historical development (the Rochelle salt period) ferroelectricity was considered an academic curiosity with no practical applications. There was little theoretical interest due to the quality of the ferroelectric materials (very fragile and water-soluble) existing at that time. The discovery of ferroelectricity in robust ceramic materials (barium titanate) during World War II launched a new era of rapid progress in the field. The structural simplicity of barium titanate stimulated numerous theoretical works, while its physical properties were utilized in many devices. Since that time, ferroelectric response has been found in a wide range of materials, including inorganic, organic, and biological species. According to [3] there are 72 families of ferroelectrics presented in Landolt–Börnstein-Vol.III/36 (LB III/36). Forty-nine of these families are inorganic crystals (19 families of oxides + 30 families of crystals other than oxides), and 23 families are organic crystals, liquid crystals, and polymers.

The enormously broad range of materials exhibiting ferroelectricity and the variety of their physical properties result in numerous applications of bulk ferroelectrics [4]. Table 1 shows the connections between different physical effects exhibited by bulk ferroelectrics and their applications.

Recent advances in nanotechnologies, especially in nanoinstrumentation (for example, scanning probe microscopy [5]) and materials nanofabrication [6], allowed the direct probing of ferroelectricity at the nanoscale. The new and unexplored world of nanoscale ferroelectrics (nanoparticles of different shapes and sizes, nanofilms, nanopatterned structures, etc.) raised fundamental questions and stimulated very active research in both academic and industrial sectors [7]. As a result, a new era of nanoscale ferroelectrics was launched. Novel effects, associated with reduced dimensions and found in nanoscaleferroelectrics, highlighted exciting possibilities for new applications reviewed recently in [8]. Almost all of the attention for the mentioned review [8] was devoted to the *thin film* nanoscale device structures (which can be easily integrated with a Si chip) with focus on ultrafast switching, electrocaloric coolers for computers, phase-array radar, three-dimensional trenched capacitors for dynamic random access memories, room temperature magnetic field detectors, and miniature X-ray and neutron sources. So far, we have not found a coherent review summarizing the actual and possible applications of ferroelectric *nanoparticles*. Our book chapter is an attempt to describe and analyze the state of the field of applications of ferroelectric nanoparticles with focus on materials technologies, medicine, and biology.

Physical effect/property	Applications
Ferroelectric hysteresis	Nonvolatile computer information storage
High relative permittivities (several thousands)	Capacitors
Direct piezoelectric effect	Sensors (microphones, accelerometers, hydrophones, etc.)
Converse piezoelectric effect	Actuators, ultrasonic generators, resonators, filters
Pyroelectric effect	Uncooled infra-red detectors
Electro-optic effects	Laser Q-switches, optical shutters and integrated optical (photonic) devices
Nonlinear optical effects	Laser frequency doubling, optical mixing, including four-wave mixing and holographic information storage
Coupling between stress and birefringence	Radar signal processing
Positive temperature coefficient of resistance (PTCR)	Electric-motor overload-protection devices and self-stabilizing ceramic heating elements

Table 1. Applications of bulk ferroelectrics

2. Methods of ferroelectric nanoparticles production: Nano-powders and nano-colloids

The existing methods that produce ferroelectric nanoparticles are numerous and can be classified as physical, chemical, and biological (Fig. 1). The primary goal of each method is to control size, shape, morphology, and crystallinity of nanoparticles to produce a desirable effect. This task represents a real challenge, and, as a result, there are no well-defined boundaries between physical, chemical, or biological methods. Moreover, in many cases, a combination of at least two methods (for example, physical and chemical methods, i.e. a physical-chemical approach) is required in order to fabricate good quality particles (small sizes, narrow size distribution, ferroelectric phase).

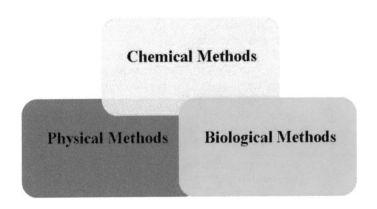

Figure 1. Methods used to fabricate ferroelectric nanoparticles

2.1. Chemical methods

The most widely used chemical methods for the synthesis of ferroelectric nanoparticles are: 1) solid-state reaction; 2) sol-gel technique; 3) solvothermal method; 4) hydrothermal method; and 5) molten salt method.

Table 2 shows selected examples of these methods applied to synthesis of ferroelectric nanoparticles of $BaTiO_3$. As can be seen from Table 2, the ultra-fine ferroelectric nanoparticles (<10 nm) in almost all cases except [16] are synthesized in a cubic phase which is not ferroelectric. The tetragonal phase (with ferroelectric response) is possible for relatively large particles (~50-70 nm). This fact can be critical for certain types of applications, and will be discussed later.

Shape of nanoparticles; Ref.	Synthesis method; Raw materials	Relevant parameters (size, morphology, crystallinity, spontaneous polarization etc.); Measurement and characterization methods	Possible applications & comments
Nanotubes; [10]	Wet chemical route at low temperature (50°C); H_2TiO_3 nanotubes; ethanol/water mixture with 25% ethanol by volume; $Ba(OH)_2 \cdot 8H_2O$	Uniform $BaTiO_3$ nanotubes, a *cubic phase*, average diameter = 10 nm and wall thickness = 3 nm at room temperature; Powder X-ray diffraction (XRD), field-emission scanning electron microscopy (FE-SEM), transmission electron microscopy (TEM), Raman spectroscopy, and X-ray photoelectron spectroscopy	Promising microwave-absorbing materials
Nanocrystals; [11]	Sol-gel technique; Barium titanium ethyl hexano-isopropoxide, $BaTi(O_2CC_7H_{15})$ $(OCH(CH_3)_2)_5$; a mixture of diphenyl ether, $(C_5H_5)_2O$; stabilizing agent oleic acid, $CH_3(CH_2)_7CHdCH(CH_2)_7-CO_2H$, at 140 °C, under argon or nitrogen	Monodisperse nanoparticles with diameters ranging from 6 to 12 nm, *cubic phase;* XRD,TEM	Multilayer ceramic capacitors
Dense polycrystalline aggregates (~80 nm) of nanocrystals (~30 nm); [12]	Solid-state reaction as a function of temperature (400–8000C), time (1–24 hr); Nanocrystalline TiO_2, ultrafine $BaCO_3$ and submicrometer $BaCO_3$ were intensively mixed in an aqueous suspension for 24 hr using polyethylene jars and zirconia media. The polymer (ammonium polyacrylate) was required for the formation of a monolayer on the particle surface	Nanoparticle size is 70 nm. Specific surface area up to ~15m^2/g, *tetragonal phase;* TG and DTA analysis, XRD, SEM	Multilayer ceramic capacitors
Polyhedral with hexagonal outline in shape; [13]	Molten salt method; Hydroxide octahydrate $(Ba(OH)_2 \cdot 8H_2O)$, titanium dioxide (TiO_2), and the eutectic salts $(NaCl–KCl)$, 600-900 °C	50 nm, *cubic phase*; XRD, Fourier transform infrared spectrometry, UV–Vis diffuse reflectance spectra, and field emission SEM	Nanoparticles are well dispersed

Shape of nanoparticles; Ref.	Synthesis method; Raw materials	Relevant parameters (size, morphology, crystallinity, spontaneous polarization etc.); Measurement and characterization methods	Possible applications & comments
Single crystalline nanoparticles; [14]	Solvothermal method; Ba(CH$_3$COO)$_2$ and Ti(OC$_4$H$_9$)$_4$; the autoclave was maintained at 200 °C for 12 hr	5-20 nm, *cubic phase*; XRD, TEM, SAED, FTIR	Dense bulk nanocomposites
Nanoparticles; [15]	Hydrothermal method; (Ba,Sr)(OiPr)$_2$, Ti(OiPr), EtOH, 330–400 °C, 16-30 MPa	15-50 nm, *cubic or tetragonal phase*; XRD, TEM	Dense bulk nanocomposites
Nanopowders; [9]	Combined wet-chemical and rapid calcination process; BaCO$_3$–TiO$_2$ precursors	125 nm, *tetragonal phase*, remnant polarization P$_r$ = 1.64 μC/cm^2, the coercive field, E$_c$ = 4.91 kV/cm; (cigar-like loop);surface area 7.96 m^2/g XRD, TEM, TGA, SMPS, SAED,dielectric spectroscopy	Multilayer ceramic capacitor
Nanocrystals; [16]	One-step solvothermal route; Tetra-n-butyl titanate, Barium hydroxide octahydrate, solvent (Diethylene glycol), surfactant (Polyvinyl Pyrrolidone)	5 nm, *tetragonal phase*; XRD, TEM, HRTEM, Raman spectroscopy	Multilayer ceramic capacitor

Table 2. Chemical methods

2.2. Physical methods

Dry and wet mechanical grindings are methods of choice for inexpensive nanoparticle preparation. For technological applications, wet grinding is preferable because it allows more options to control the size of the nanoparticles. During the last few years, substantial progress was made in the field of ferroelectric nanoparticle preparation by means of wet mechanical grinding [17]. Generally, three components are needed: raw material to grind (micron-sized powders of BaTiO$_3$); surfactant (which covers the particle surface and prevents their aggregation and overheating during grinding; oleic acid is a good choice for BaTiO$_3$); and fluid carrier (both raw material and surfactant are mixed with fluid carrier; heptane is widely used to grind BaTiO$_3$). An extensive list of references, as well as recent achievements in the field, are discussed in the review [18]. Table 3 shows how particle sizes depend upon grinding time [19].

As is seen from Table 3, wet mechanical grinding can produce 9 nm nanoparticles, and such small particles are still in ferroelectric phase. It is important to remember that BaTiO$_3$ nanoparticles of the same sizes synthesized by the majority of chemical methods are not ferroelectric – the method used to fabricate nanoparticles really does matter. The ability to produce ferroelectric nanoparticles of very small sizes is a defining characteristic of this physical method (see [18] for more references).

Grinding time, hrs	3	7	10	16	20	24
Average particle diameter, nm	26	14.9	11.6	9.5	9.2	9

Table 3. BaTiO$_3$ particle sizes as a function of grinding time (liquid carrier – heptane; two-station PM200 planetary ball mill from Retsch)

The net dipole moment of ferroelectric nanoparticles allows them to be harvested by using an inhomogeneous electric field. The harvesting concept proposed in [20] is based on the fact that dipoles experience a translational force only when exposed to a field gradient. For a given linear field gradient and assuming a single ferroelectric domain, the net translational force on a dipole scales proportionally with the particle size. The Brownian motion effects also become progressively more pronounced at smaller particle sizes and so the required field strength for successful separation scales nonlinearly as the particle size is reduced.

Both gas-phase and liquid-phase harvesting methods were proposed and tested successfully, and single ferroelectric monodomain nanoparticles as small as 9 nm from mechanically ground nanoparticles were selectively harvested [20]. In contrast to many reports on the lack of ferroelectricity for nanoparticles below 10 nm (see Table 2), the harvested nanoparticles do maintain ferroelectricity. The ferroelectric response of such tiny nanoparticles was attributed to the existence of an induced surface strain as a result of the grinding process [20]. The lack of a mechanically induced strain in similarly sized but chemically produced nanoparticles accounts for the absence of ferroelectricity in these materials. The concept of stress-induced ferroelectricity was verified in further experiments, which are reviewed in [18]. It was found [21] that the spontaneous polarization of the nanoparticles is four to five times larger than the spontaneous polarization of the bulk raw materials. To obtain this result the following two conditions must be satisfied: (1) a nonpolar solvent for the nanoparticle suspension and (2) nonaggregated nanoparticles. Under these conditions, for 9 nm nanoparticles, the values of 100–120 $\mu C/cm^2$ and 8.9 $10^{-23}C$ cm have been measured for the spontaneous polarization and dipole moment, respectively. The aggregation of ferroelectric nanoparticles masks the ferroelectric response due to the partial compensation of the dipole moments of the individual particles. Finally, we can conclude that recent advances in the production of uniform, monodomain, highly ferroelectric nanoparticles indicate that this field has reached its maturity. The particles can now be reliably prepared to certain specifications and characteristics [18].

2.3. Physical-chemical methods

In order to prepare ferroelectric nanoparticles with controllable sizes and shapes a combination of chemical methods with external physical factors (for example, a chemical reaction in the presence of electromagnetic fields or mechanical milling) is needed. Table 4 shows several examples of these physical-chemical methods for the case of BaTiO$_3$.

As can be seen from Table 4, chemical reactions under the presence of an external driver (microwave, ultrasonic, milling, heat, pressure) are able to produce very fine particles (~5-10 nm) with tetragonal structures.

Shape of nanoparticles; Ref.	Synthesis method; Raw materials	Relevant parameters (size, morphology, crystallinity, spontaneous polarization etc.); Measurement and characterization methods	Possible applications & comments
Truncated nanocubes; [22]	Microwave (2.45 GHz, output power ~ 800W) hydrothermal; High-purity reagents barium nitrate (Sigma Aldrich, purity ≥99.99%), titanium isoproxide (Sigma Aldrich, purity ≥97%), nitric acid (ACS reagent 70%), ammonium hydroxide (Sigma Aldrich), and glycine (Sigma-Aldrich, purity ≥99%)	Cross section 70 ± 9 nm in [100] projection; *tetragonal structure*; remnant polarization is 15.5 $\mu C/cm^2$; saturation polarization is 19.3 $\mu C/cm^2$; XRD, high resolution TEM, impedance spectroscopy	Charge storage devices
Nearly spherical nanoparticles; [23]	Sol–gel-hydrothermal method under an oxygen (partial pressure is ~60 bar); $TiCl_4$; HCl; $BaCl_2 \cdot 2H_2O$; deionized water, stirring and N_2 bubbling, NaOH	Sizes range from 50 to 75nm; transition of $BaTiO_3$ from *cubic* to a *pseudotetragonal phase* with the increase ofthe reaction temperature from 80 to 220 °C; remnant polarization $P_r = 2.2$ C/cm^2; the coercive field, $E_c = 3.2$ kV/cm (cigar-like loop); XRD, TEM, Raman and dielectric spectroscopy; energy-dispersive X-ray spectroscopy; AFM	Oxygen atmosphere decreases sizes of the nanoparticles (for example , at 200 °C sizes decrease from 72.54 nm to 49.54 nm)
Nanocrystals; [24, 25]	Direct synthesis from solution (DSS) – mechanochemical synthesis; Anhydrous $Ba(OH)_2$ and tetrabutyl titanate [Ti $(OC_4H_9)_4$] Mechanical milling at the rate of 200 rpm, room temperature	7 nm, *tetragonal structure*; XRD, TEM, Raman spectroscopy	Chemical reaction during milling
Large aggregates of 5–10 nm small nanocrystals; [26]	Sonochemical synthesis; $BaCl_2$; $TiCl_4$; NaOH	100 nm aggregates, *cubic phase*; XRD, TEM, DLS, SAED	Multilayer ceramic capacitors

Shape of nanoparticles; Ref.	Synthesis method; Raw materials	Relevant parameters (size, morphology, crystallinity, spontaneous polarization etc.); Measurement and characterization methods	Possible applications & comments
Single nanocrystals; [27]	RF-Plasma chemical vapor deposition (CVD); iso-propoxide (Ti(OiPr)₄, bis-dipivaloylmethanate (Ba(DPM)₂	12.2-15.4 nm; XRD, TEM, SAED, EDX	Multilayer ceramic capacitors
Nanofibers; [28, 29, 30]	Electrospinning, sol-gel; barium acetate, PVP, titanium isopropoxide, ethanol, acetic acid	80-190 nm in diameter, 0.1 mm in length; XRD, TEM, SAED	Piezo, microwave material, sensors, capacitors
Nanopowders; [31]	Flame spray pyrolysis; barium carbonate and titanium tetra-iso-propoxide, citric acid	70 nm, *tetragonal structure*; XRD, TEM, SEM	Multilayer ceramic capacitors

Table 4. Physical-chemical methods

Another interesting subject is the possible application of ceramic nanofibers produced by the method of electrospinning. These possible applications include: nanofiber-based supports for catalysts, nanofiber-based photocatalysts, nanofiber-based membrane for filtration, nanofiber-based sensors, nanofiber-based photoelectrodes for photovoltaic cells, nanofiber-based electrodes for lithium batteries, nanofiber-based electrode-supports for fuel cells, and materials for implants [28, 29, 30].

2.4. Biological methods

The proper combination of chemical and physical methods (Table 4) allows the production of very small ferroelectric nanoparticles (5-10 nm). However, toxic chemicals and/or high temperatures/pressures are needed in most cases. These requirements substantially limit the possible biomedical applications of ferroelectric nanoparticles. Biological methods were proposed as eco-friendly "green" alternatives to existing chemical and physical methods. The biosynthesis of different types of nanoparticles was reviewed recently in a few papers [32, 33, 34, 35], but biological methods that produce specifically ferroelectric nanomaterials are not very numerous. For example, the first review [32] lists nearly a hundred variations for the synthesis of nanoparticles by microorganisms, and only a few of them produce ferroelectric nanoparticles. The biological methods applied to synthesize the nanoparticles of the most widely known ferroelectric $BaTiO_3$ are summarized in Table 5 (papers published before 2008 were discussed in [36]). The references listed in this table show that most of the biological methods mentioned employ some chemical or physical steps as well.

Shape of nanoparticles; Ref.	Synthesis method; Raw materials	Relevant parameters (size, morphology, crystallinity, spontaneous polarization etc.); Measurement and characterization methods	Comments
Aggregates (~500 nm) of nanocrystals(~ 50 nm); [37]	The peptides BT1 and BT2induce the room-temperature *precipitation* of $BaTiO_3$; Aqueous precursor solution composed of barium acetate $(Ba(OOCCH_3)_2)$ potassium bis(oxalato) oxotitanate(IV) $(K_2[TiO(C_2O_4)_2]2H_2O)$; pH 6.8	50-100 nm; *tetragonal phase* remnant polarization $P_r \sim$ 2-3 C/cm²; the coercive field, E_c = 5-6 kV/cm (cigar-like loop); SEM, TEM, SAED and XRD analysis, dielectric measurements	Rapid room-temperature synthesis of ferroelectric (tetragonal) $BaTiO_3$ within 2 hr
Quasi-spherical nanoparticles; [38]	*Lactobacillus* - assisted biosynthesis; $BaCO_3$ and TiO_2 solid state synthesis $BaTiO_3$; Slush of micron-sized $BaTiO_3$ particles+ Lactic acid Bacillus spore tablets	20-80 nm, *single-phase tetragonal structure*; XRD, TEM	Extracellular synthesis
Spherical nanoparticles; [39]	*(Fusarium oxysporum)* fungus-assisted biosynthesis of $BaTiO_3$ nanoparticles; $(CH_3COO)_2Ba$ and K_2TiF_6 + fungal micelles of *Fusarium oxysporum*	4-5 nm, *tetragonal structure*; XRD, DCS, TEM, SAED, XPS, SPM	Extracellular synthesis
Spherical nanoparticles; [40]	Peptide nanorings used as templates; $BaTi(O_2CC_7-H_{15})[OCH(CH_3)_2]_5$	6-12 nm, *tetragonal structure*; AFM, TEM, XRD, EFM, Raman	4 days
Spherical nanoparticles; [41]	*(Saccharomyces cerevisiae)* baker's yeast-assisted biosynthesis of $BaTiO_3$ nanoparticles; $BaCO_3$ and TiO_2 solid state synthesis $BaTiO_3$, Slurry of large $BaTiO_3$ + yeast culture	8-21 nm, *single-phase hexagonal structure*; XRD, TEM	Extracellular synthesis
Spherical nanoparticles; [42]	Kinetically controlled vapor diffusion Single source, bimetallic alkoxide with the vapor diffusion of a hydrolytic catalyst (H_2O)	6-8 nm, *cubic phase*; XRD, TEM	~250 g.

Table 5. Biological methods

One can note from Table 5 that these biological methods can be roughly grouped into two categories: the synthesis of ferroelectric nanoparticles using a number of seed chemicals and some type of biological or bio-inspired system, and the creation of nanoparticles using large particles of ready ferroelectric material and some type of biological system. Most of the methods that are currently available fall into the first category; the methods in the second category have appeared only very recently.

The earliest discoveries of nanoparticle synthesis by microorganisms (fungus, *Lactobacillus* and yeast) involve metal, alloy, and metal oxide nanoparticles. With a lag of one to two years, this research was followed by attempts at ferroelectric nanoparticle synthesis by the same living systems [32]. Fungus (*Fusarium oxysporum*, commonly found in soil) has been shown to synthesize extracellularly ferroelectric nanoparticles of barium titanate (4-5 nm of average size) [39, 43] at room temperature, producing small ferroelectric nanoparticles on a scale that has been previously inaccessible. This method falls into the first category, since fungus produces barium titanate from more than one seed chemical. *Lactobacillus* (bacteria commonly used to curdle milk and produce buttermilk) has been shown to synthesize ferroelectric nanoparticles (with sizes ranging from 20 to 80 nm) from the slush of large barium titanate particles [38]. Baker's yeast (*Saccharomyces cerevisiae*) has also been attempted as a biosynthesis agent; the barium titanate nanoparticles received were on average ~10nm [41]. Both the yeast- and lactobacillus-mediated production of barium titanate nanoparticles fall into the second category: they start with large particles of ready barium titanate in water and produce small (under 100 nm) nanoparticles extracellularly.

Peptide nanorings provide another interesting biomimetic route to the template-mediated synthesis of $BaTiO_3$ and $SrTiO_3$ nanoparticles [40]. Developed in 2006, it was the very first method of obtaining ferroelectric barium titanate nanoparticles at room temperature (many previous room-temperature methods had trouble growing barium titanate not in the cubic phase).

In this category of biological methods, one has to mention bioinspired methods of dispersing nanoparticles and producing stable colloids. Take for example the production of stable nanoparticle suspensions using microbial-derived surfactants [44] where usual garden variety surfactants are replaced with bio-derived materials that are significantly smaller in size and that help prevent aggregation of multiple nanoparticles during sol-gel synthesis. This process relies on some physical/chemical steps, just like the majority of other biological methods of ferroelectric nanoparticle production that have been developed to date.

These bio-inspired methods show great promise because they produce relatively small ferroelectric nanoparticles at room (or low) temperature compared to conventional physical and chemical methods, many of which require lengthy processes, use of high temperatures, harsh chemicals, etc.

3. Applications of ferroelectric nanoparticles in materials technologies

3.1. Multilayered capacitors and nanocomposites

As was previously mentioned in the introduction, the era of applied ferroelectricity was launched after the first reliable ferroelectric material – ceramic $BaTiO_3$ – was discovered [1]. The very high dielectric constants (~1000) of bulk ferroelectric materials were applied in the manufacturing of discrete and multilayered ceramic capacitors. Advances in the production of ferroelectric nanoparticles broadened this application to a large extent. Currently, nanosized $BaTiO_3$ powders are successfully used for manufacturing miniaturized and highly volume-efficient multilayer ceramic capacitors (MLCCs, for more details see review [45]).

Typical MLCCs utilize pellets of nanopowders sintered at high temperatures, producing structures with high rigidity that can impose certain limitations. Nanocomposites made of flexible polymers and ferroelectric nanoparticles overcome the restrictions caused by rigidity of sintered structures. Polymeric materials doped with ferroelectric nanoparticles are of considerable interest as a solution for processable high permittivity materials for various electronic applications: volume efficient multilayer capacitors, high-energy-density capacitors, embedded capacitors, and gate insulators in organic field effect transistors [46].

Both components (polymer and ferroelectric nanoparticle) of such nanocomposites are essential. The polymeric material brings flexibility and the ability to cover a large surface area, while nanoparticles share high values of dielectric permittivity with the matrix-enhancing effective dielectric permittivity of the composite. It's not easy to achieve homogeneous dispersion of ferroelectric nanoparticles in a polymeric matrix because of the high surface energy of nanoparticles, which usually leads to aggregation and phase separation, resulting in poor processability of the films and a high defect density. In order to make high-quality nanocomposites (good solution processability, low leakage current, high permittivity, and high dielectric strength), surface modification of nanoparticles is needed. The surface modification of nanoparticles decreases attractive forces between them, thus preventing aggregation. In addition, the proper design of a surface agent can increase interactions between the nanoparticles and the polymeric matrix. As a result, a quite homogeneous distribution of the nanoparticles in the polymeric matrix can be achieved. However, there are several obstacles that prevent nanocomposites "polymer/ferroelectric nanoparticle" from mass commercialization. The most common are poor temperature stability of dielectric constants and large dielectric losses. The proper design of both the nanoparticles (material size, shape, surface modification) and the polymer matrix can overcome these problems. For example, Table 6 shows short descriptions of the relatively successful (in terms of material performance) nanocomposites "polymer/BaTiO$_3$". All of them are optically transparent, have high dielectric permittivity (25-30), and moderate losses (loss tangent is in the range 0.01-0.05 for different materials).

Composition (polymer/ nanoparticle)	Particle surface functionalization	Dielectric permitti-vity of nanocom-posite	Particle size &breakdown electric field	Ref.
Barium titanate / poly(vinylidene fluoride-co-hexafluoro propylene), volume concentration < 80%	Phosphonic acid surface-modified BaTiO₃ nanoparticles	35	30-50 nm >164 V/µm	[46]
Barium titanate/polyimide (BaTiO₃/PI); volume concentration < 40%	Core-shell structure; polyamic acid (PAA) was used to cover particle	20	70-100 nm 67 MV/m	[47]
Poly(vinylidene fluoride)/ Barium titanate (PVDF/ BaTiO₃); volume concentration 10%- 30%	Surface hydroxylated BaTiO₃	18-25	85-100 nm	[48]
Barium titanate/polyimide (BaTiO₃/PI); volume concentration < 50%	BaTiO₃ nanoparticles were prepared by the alkoxide route	30	30-50 nm 50-80 nm	[49]
Barium titanate/poly(methyl methacrylate) (BaTiO₃/PMMA); volume concentration 53%	BaTiO₃ nanoparticles were modified with a silane coupling agent (3-methacryloxypropyltrimethoxysilane)	36	16-21 nm	[50]
Barium titanate/polyimide (BaTiO₃/PI); volume concentration 59%	BaTiO₃ nanoparticles are surface modified by phthalimide with the aid of a silane coupling agent as a scaffold	37	20 nm	[51]

Table 6. Nanocomposites of polymer/ferroelectric nanoparticles

3.2. Ferroelectric nanoparticles and liquid crystals: Display and non-display applications

It is an established fact that modern technologies require novel and highly advanced materials. During the last decade, nanochemistry has developed a dozen conceptually different pathways of synthesis to satisfy the constantly growing demand for new materials [52]. However, chemical methods are elaborate, time consuming, and expensive. Although they are generally accepted as universal, nanochemistry methods do not work perfectly in certain cases. For example, it's very difficult (but not impossible) to produce 5-10 nm ferroelectric nanoparticles with tetragonal structures through chemical methods (see Table 2). In many cases, the chemical methods used to produce novel materials can be efficiently supplemented or even replaced with non-synthetic ones. The addition of ferroelectric nanoparticles to different materials (polymers; see Table 6, liquids, liquid crystals etc.) is a good example of a non-synthetic method used to develop materials with improved properties. This subsection will discuss how ferroelectric nanoparticles can modify properties of liquid crystals – materials that found numerous applications in the display industry, photonics, optical processing, and the biotech industry.

The concept of ferroelectric colloids in liquid crystals was created by Yu. Reznikov and co-authors [53] – liquid crystal material was doped with stabilized ferroelectric nanoparticles of low concentration (~0.3 %). Ferroelectric nanoparticles share their intrinsic properties with the liquid crystals matrix due to the alignment within the liquid crystal and interactions between mesogenic molecules. The low concentration of nanoparticles and their stabilization with surfactant (oleic acid) provided the stability of such systems. This classic paper reported on the main features of such colloids: (1) a lower threshold voltage by a factor of 1.7; (2) an enhanced dielectric anisotropy by a factor of 2; (3) a linear electro-optical response (the sensitivity to the sign of an applied electric field). It should be pointed out that pure nematics are not sensitive to the sign of an electric field – this property is intrinsic to ferroelectric liquid crystals rather than to nematics. The first experimental results stimulated a very active global interest in the field. More than 100 papers were published during the last decade, and, in the last few years, several review papers summarized the most important results [18, 54, 55]. The review paper [54] published this year by the founder of this field is of special interest since it comprehensively discusses the past, the present, and the future of the liquid crystal colloids.

Ferroelectric nanoparticles embedded into a liquid crystal host strongly interact with the surrounding mesogenic molecules, due to the strong permanent electric field of this particle. These interactions can affect the basic physical parameters of liquid crystals: birefringence, dielectric permittivity, elastic constants, viscosity, electrical and thermal conductivity, temperatures of phase transitions, etc. There are two basic mechanisms responsible for the observed effects: (1) the increase of the orientation coupling between mesogenic molecules; and (2) the direct contribution of the permanent polarization of the particle [54]. Experimental data suggests that the first scenario (the increase of the orientation coupling between mesogenic molecules) is more likely to happen in the case of single component liquid crystals [54]. The second mechanism (associated with the direct contribution of the permanent polarization of the ferroelectric nanoparticle to the physical properties of liquid crystals) is the primary factor in the case of multi-component liquid crystal mixtures [54].

So far most of the reported experimental data describes the properties of multi-component nematics: liquid crystals doped with ferroelectric nanoparticles. The electric field generated by ferroelectric nanoparticles can cause micro- and/or nano-separation of such mixtures and affect the macroscopic properties of the liquid crystal colloid. As a result, in some cases, the data found for single component liquid crystals (5CB) is different from the data obtained for multi-component mixtures. Nevertheless, an analysis of the existing literature [18] shows that most of the published experimental data report that ferroelectric nanoparticles embedded in liquid crystals at low concentrations: (1) enhance dielectric permittivity, dielectric anisotropy, and optical birefringence; (2) lower the switching voltage U_{th} for the Freedericksz transition; (3) increase the orientational order parameter S and the isotropic–nematic transition temperature T_{NI}; and (4) reduce the switching times needed to reorient liquid crystals by an external electric field. It should be pointed out that in all of these cases, wet grinding was used to prepare the ferroelectric nanoparticles. Few papers claim results opposite to the ones (1)-(4) listed above; however, nanoparticles used to make a liquid crystal colloid in this case were not ferroelectric [18]. The important conclusion to take from these findings is that a strong ferroelectric response of ferroelectric nanoparticles is a key factor leading to all "positive" effects (1)-(4).

The stability of liquid crystal colloids is the most challenging problem that prevents the transition of liquid crystal colloids of ferroelectric nanoparticles from the academic sector to the industrial domain. During the last few years, the general focus of the research has shifted toward this problem, and the proper surface functionalization of nanoparticles is now being considered as one of the most important factors affecting the stability of liquid crystal colloids. The shape and size of nanoparticles are also of the utmost importance – it was found that 10-20 nm ferroelectric nanoparticles affect the properties of a liquid crystal host much morethan the same but larger nanoparticles(~100 nm).

In summary, ferroelectric nanoparticles can modify the intrinsic properties of liquid crystal materials without time-consuming and expensive chemical synthesis. Experimental data on the enhancement of electro-optical, optical, and nonlinear-optical responses of such materials shows the strong potential of ferroelectric nanoparticles for improving the "practical" properties of liquid crystals, especially for those materials where the method of chemical synthesis has reached its limit. Such modified materials are very attractive and suitable for use in switchable lenses, displays, and beam steering, as well as other light-controlling devices (spatial light modulators, tunable filters etc.) [18].

4. Ferroelectric nanoparticles in biology

In the last decade, ferroelectric nanoparticle areas of application veered toward biology and medicine. The first challenge to address when introducing nanoparticles to biological and medical systems is how to make them stable in media that is bio-compatible, namely in aqueous solutions. If the ferroelectric nanoparticles are not coated with a stabilizing agent, the intrinsic properties of such nanoparticles very often lead to unwanted effects such as aggregation and precipitation, as well as the leaching of some ions (leading to change in particle properties) in aqueous solutions. There has been a tremendous amount of effort in the last twenty years to create many species of ferroelectric nanoparticles with different surface coatings [56and references within]. For example, such stabilizing agents as polyacrylic acid, polymethacrylic acid, polyaspartic acid, (aminomethyl) phosphonic acid and poly-l-lysine were tried on barium titanate nanoparticles. A large variety of possible coatings yields different surface properties of said nanoparticles – different thickness and charge of the surface layer, different strength of stabilizing agent adhesion, different resulting particle size – all of it leading to very different interactions with biological objects such as cells and cellular components. Such water-soluble ferroelectric nanoparticles can be further functionalized with fluorescent markers or antibodies [57], and they have been recently observed within a number of mammalian cells [58, 59, 60] in vitro. The majority of ferroelectric nanoparticles that have been successfully functionalized are in the middle of nano size range (~100 nm or more), and the surface functionalization very frequently adds a thick surface layer, resulting in particles that are a few hundred nanometers in size. This poses a challenge which a few researchers are attempting to overcome through bio-inspired surfactants [44]. The surface functionalization of nanoparticles can be crucial for a desired effect in a biosystem – the response is often cell-specific, and biocompatibilty has to be experimentally tested for each new combination of nanoparticle and cell [61]. For example, a recent study [62] indicates that larger (over 200 nm) barium titanate nanoparticles have been successfully used with and without surface modification and they cause no toxic response in endothelial cell cultures.

Current efforts in this area concentrate on carrying these findings through to the systems of cells and tissue in vivo [63, 64]. Such biocompatible nanoparticles can be used in a variety of ways: as imaging agents [58, 59, 60, 64, 65], biocompatible nanoprobes [62, 66, 67, 68], cell proliferation agents [65], etc. Ferroelectric nanoparticles are currently being widely explored for applications in the area of medical imaging, e.g. for enhancing contrast in second harmonic generating methods for imaging deep tissue in vitro and in vivo [64, 69] for such important applications as screening for genetic diseases or cancer[70].

5. Ferroelectric nanoparticles applications in medicine and medical engineering

As mentioned above, ferroelectric nanoparticles have become widely used in material science and engineering applications, but biological and medical applications of these fascinating nanoparticles has only begun to be explored in the last decade. It has been recently discovered that they can be used for cell imaging or the detection of malignancies in lung cancer studies, or they can be functionalized to induce cell proliferation. Ferroelectric materials have a non-centrosymmetric crystalline structure, and are thus capable of generating a second harmonic of light [67]. This distinctive feature of ferroelectrics is the basis for a growing number of applications of ferroelectric nanoparticles as imaging/diagnostic agents and nanoprobes in optical imaging [64].

Optical imaging and detection methods are the most widespread among biological and medical communities. For example, second harmonic generation imaging has been successfully used for detection of *osteogenesis imperfecta* in biopsies of human skin [71] and lung cancer [72]. To improve contrast, many of the imaging methods rely on imaging probes, such as fluorescent markers or quantum dots [73]. This is the case for the SHG imaging technique: the SHG signal from biological cell components is often weak, so novel nanoprobes have been introduced to enhance contrast. SHG probes are photo-stable, and do not bleach or blink unlike conventional fluorescent probes. By definition, second harmonic generating nanoprobes (such as ferroelectric nanoparticles) are capable of converting two photons of light into one photon of half the incident wavelength [74]. This second-harmonic light can be detected using methods of nonlinear optical microscopy. Nonlinear optical properties of ferroelectric nanomaterials can be used for optical phase conjugation [75] and nonlinear microscopy [62, 76] – these properties have allowed them to spread to the area of medical sensors.

This also gives the ferroelectric SHG particle an edge in cell and tissue imaging in vivo. For example, recent advances in this area include imaging a tail of a living mouse with the aid of barium titanate nanoparticles [69].

The intrinsic large values of the dielectric permittivities of ferroelectric nanoparticles suggest their use to enhance the dielectric contrast of materials, such as polymers [77 and references within] and biological tissue [70]. These unique properties of ferroelectric nanoparticles lead to their novel use as contrast-enhancing agents for microwave tomography, which is a method of non-invasive assessment and diagnostics of soft tissues (such as detecting malignancies)[70].Recently, ferroelectric electro spun nanofibers also emerged in various biomedical areas including medical prostheses, tissue engineering, wound dressing, and drug delivery [28,29,30].

In conclusion, ferroelectric materials found a wide variety of biomedical applications in the last decade – and the list is constantly growing [78]. The ferroelectric material (e.g. barium titanate) used in medical implants has been known to accelerate osteogenesis [79], and the same material in nanoparticle form works both as an SHG probe to detect Osteogenesis Imperfecta [71] and, through microwave tomography, to detect lung cancer [70]. We expect more applications will become possible if other effects, such as piezoelectricity, ferroelectric hysteresis or stress-birefringence coupling, are exploited with biology and medicine in mind.

6. Conclusions

A review of recently published research papers allows us to conclude that the unique properties of ferroelectric nanoparticles offer an enormous range of applications, especially in material technologies, biology, and medicine. However, since such a conclusion has become an academic cliché, we would like to make just a few realistic and optimistic comments on the subject.

Discussing any application, we have to be more specific and distinguish between potential applications and actual (i.e. mass commercialized) ones. So far, the "applied" field of ferroelectric nanoparticles is at the stage of high potential for commercialization rather than real mass-commercialization. However, after the very first proof-of-concept research studies were completed just recently, a serious shift toward real applications was initiated. The research community realized that the controllable preparation of nanoparticles (size, shape), their proper surface functionalization, and the homogeneous dispersion into host material (liquid crystals, polymers, biological species) are among the most critical steps on the way to mass-commercialization. Many applications (especially bio-medical) require a large-scale preparation of mono-dispersed, very fine (5-10 nm) ferroelectric nanoparticles in the tetragonal phase – and only recently has substantial progress been made in this direction (see section 2). However, such ultra-fine nanoparticles were mostly tested in material technologies (ferroelectric liquid crystal colloids), while bio-medical methods utilized relatively large ferroelectric nanoparticles (~50-100 nm) (sections 4-5). The bioconjugation of 5-10 nm ferroelectric nanoparticles can extend the area of biomedical application even more, since these nanoparticles can be used as both passive and active nanoprobes.

The applied science of ferroelectric nanoparticles is in the beginning of its development. Many fundamental and applied problems still need to be solved before the potential applications will be converted into actual ones. Nevertheless, despite the fact that the emerging problems in applying nanoparticles in material technologies and biomedicine are still numerous [80], we can conclude that a good balance between purely academic and purely applied foci of research is the key toward mass commercialization of ferroelectric nanoparticles in all applied fields mentioned above.

Acknowledgements

The authors are very grateful to all coauthors of the papers related to the topic of ferroelectric nanoparticles. We acknowledge also the support of the UCCS Center for the Biofrontiers Institute, University of Colorado at Colorado Springs. A great portion of the work described in this review has been supported by the grants from the NSF #1102332 "Liquid Crystal Signal Processing Devices for Microwave and Millimeter Wave Operation" and # 1010508 "STTR Phase 1: Design, Fabrication, and Characterization of Ferroelectric Nanoparticle Doped Liquid Crystal / Polymer Composites."

List of abbreviations

AFM atomic force microscopy

DCS differential scanning calorimetry

DLS dynamic light scattering

DTA differential thermal analysis

EDX energy dispersive X-rayspectroscopy

EFM electrostatic force microscopy

FE-SEM field-emission scanning electron microscopy

FTIR Fourier transform infrared spectroscopy

HRTEM high resolution transmission electron microscopy

MLCC multilayer ceramic capacitors

PAA polyamic acid

PTCR positive temperature coefficient of resistance

PMMA poly(methyl methacrylate)

PVDF poly(vinylidene fluoride)

SAED selected area electron diffraction

SEM scanning electron microscopy

SHG probe second harmonic generating probe

SMPS scanning mobility particle size

SPM scanning probe microscopy

TEM transmission electron microscopy

TG or TGA thermo gravimetric analysis

XPSX-ray photoelectron spectroscopy

XRDX-ray diffraction

Author details

Yuriy Garbovskiy, Olena Zribi and Anatoliy Glushchenko

UCCS Center for the Biofrontiers Institute, Department of Physics, University of Colorado at Colorado Springs, Colorado Springs, Colorado, USA

References

[1] Haertling, G.H. Ferroelectric ceramics: history and technology. *Journal of the American Ceramic Society*82, 797-818 (1999).

[2] Cross, L.E. & Newnham, R.E. in *High-Technology Ceramics—Past, Present,and Future* 289–305 (American Ceramic Society, Westerville, OH, 1987).

[3] Mitsui, T. in Springer Handbook of Condensed Matter and Materials Data (ed. Warlimont, W.M.a.H.) 903 – 938 (Springer, Berlin, 2006).

[4] Whatmore, R. in Springer Handbook of Electronic and Photonic Materials (ed. Capper, S.K.a.P.) 597 – 623 (Springer, Berlin, 2006).

[5] Kalinin, S.V. et al. Nanoscale Electromechanics of Ferroelectric and Biological Systems: A New Dimension in Scanning Probe Microscopy. *Annual Review of Materials Research*37, 189-238 (2007).

[6] Gruverman, A. & Kholkin, A. Nanoscale ferroelectrics: processing, characterization and future trends. *Reports on Progress in Physics*69, 2443-74 (2006).

[7] Ahn, C.H., Rabe, K.M. & Triscone, J.M. Ferroelectricity at the nanoscale: local polarization in oxide thin films and heterostructures. *Science*303, 488-91 (2004).

[8] Scott, J.F. Applications of modern ferroelectrics. *Science*315, 954-9 (2007).

[9] Sarkar, D. Synthesis and Properties of $BaTiO_3$ Nanopowders. *Journal of the American Ceramic Society*94, 106-110 (2011).

[10] Zhu, Y.-F., Zhang, L., Natsuki, T., Fu, Y.-Q. & Ni, Q.-Q. Facile Synthesis of $BaTiO_3$ Nanotubes and Their Microwave Absorption Properties. *ACS Applied Materials & Interfaces*4, 2101-2106 (2012).

[11] O'Brien, S., Brus, L. & Murray, C.B. Synthesis of Monodisperse Nanoparticles of Barium Titanate: Toward a Generalized Strategy of Oxide Nanoparticle Synthesis. *Journal of the American Chemical Society*123, 12085-12086 (2001).

[12] Buscaglia, M.T., Bassoli, M., Buscaglia, V. & Vormberg, R. Solid-State Synthesis of Nanocrystalline $BaTiO_3$: Reaction Kinetics and Powder Properties. *Journal of the American Ceramic Society*91, 2862-2869 (2008).

[13] Zhang, Y., Wang, L. & Xue, D. Molten salt route of well dispersive barium titanate nanoparticles. *Powder Technology*217, 629-633 (2012).

[14] Wei, X. et al. Synthesis of Highly Dispersed Barium Titanate Nanoparticles by a Novel Solvothermal Method. *Journal of the American Ceramic Society*91, 315-318 (2008).

[15] Hayashi, H. & Hakuta, Y. Hydrothermal Synthesis of Metal Oxide Nanoparticles in Supercritical Water. *Materials*3, 3794-3817 (2010).

[16] Zhang, H. et al. Fabrication of Monodispersed 5-nm $BaTiO_3$ Nanocrystals with Narrow Size Distribution via One-Step Solvothermal Route. *Journal of the American Ceramic Society*94, 3220-3222 (2011).

[17] Atkuri, H. et al. Preparation of ferroelectric nanoparticles for their use in liquid crystalline colloids. *Journal of Optics A: Pure and Applied Optics*11, 024006 (2009).

[18] Garbovskiy, Y. & Glushchenko, A. in Solid State Physics (ed. Camley, R.E.) 1 – 74 (2010).

[19] Cook, G. et al. Nanoparticle doped organic-inorganic hybridphotorefractives. *Optics Express*16, 4015-4022 (2008).

[20] Cook, G. et al. Harvesting single ferroelectric domain stressed nanoparticles for optical and ferroic applications. *Journal of Applied Physics*108, 064309-4 (2010).

[21] Basun, S.A., Cook, G., Reshetnyak, V.Y., Glushchenko, A.V. & Evans, D.R. Dipole moment and spontaneous polarization of ferroelectric nanoparticles in a nonpolar fluid suspension. *Physical Review B*84, 024105 (2011).

[22] Swaminathan, V. et al. Microwave synthesis of noncentrosymmetric $BaTiO_3$ truncated nanocubes for charge storage applications. *ACS Appl Mater Interfaces*2, 3037-42 (2010).

[23] Fuentes, S. et al. Synthesis and characterization of BaTiO₃ nanoparticles in oxygenat-mosphere. 505, 568–572 (2010).

[24] Qi, J.Q. et al. Direct synthesis of ultrafine tetragonal BaTiO₃ nanoparticles at room temperature. *Nanoscale Research Letters*6, 466 (2011).

[25] Kong, L.B., Zhang, T.S., Ma, J. & Boey, F. Progress in synthesis of ferroelectricceramicmaterialsviahigh-energymechanochemicaltechnique. 53, 207–322 (2008).

[26] Dang, F. et al. Aneweffect of ultrasonication on the formation of BaTiO₃ nanoparticles. 17, 310–314 (2010).

[27] Suzuki, K. & Kijima, K. Well-crystallized barium titanate nanoparticles prepared by plasma chemical vapor deposition. 58, 1650–1654 (2004).

[28] Ramaseshan, R., Sundarrajan, S., Jose, R. & Ramakrishna, S. Nanostructured ceramics by electrospinning. *Journal of Applied Physics*102, 111101-17 (2007).

[29] He, Y. et al. Humidity sensing properties of BaTiO₃ nanofiber prepared via electrospinning. *Sensors and Actuators B: Chemical*146, 98-102 (2010).

[30] Dai, Y., Liu, W., Formo, E., Sun, Y. & Xia, Y. Ceramic nanofibers fabricated by electrospinning and their applications in catalysis, environmental science, and energy technology. *Polymers for Advanced Technologies*22, 326-338 (2011).

[31] Jung, D.S., Hong, S.K., Cho, J.S. & Kang, Y.C. Nano-sizedbariumtitanatepowders with tetragonalcrystalstructureprepared by flamespraypyrolysis. 28, 109–115 (2008).

[32] Li, X., Xu, H., Chen, Z.-S. & Chen, G. Biosynthesis of Nanoparticles by Microorganisms and Their Applications. *Journal of Nanomaterials*2011 (2011).

[33] Chen, C.-L. & Rosi, N.L. Peptide-Based Methods for the Preparation of Nanostructured Inorganic Materials. *Angewandte Chemie International Edition*49, 1924-1942 (2010).

[34] Briggs, B.D. & Knecht, M.R. Nanotechnology Meets Biology: Peptide-based Methods for the Fabrication of Functional Materials. *The Journal of Physical Chemistry Letters*3, 405-418 (2012).

[35] Dhillon, G.S., Brar, S.K., Kaur, S. & Verma, M. Green approach for nanoparticle biosynthesis by fungi: current trends and applications. *Critical Reviews in Biotechnology*32, 49-73 (2011).

[36] Beier, C.W., Cuevas, M.A. & Brutchey, R.L. Room-Temperature Synthetic Pathways to Barium Titanate Nanocrystals. *Small*4, 2102-2106 (2008).

[37] Ahmad, G. et al. Rapid Bioenabled Formation of Ferroelectric BaTiO₃ at Room Temperature from an Aqueous Salt Solution at Near Neutral pH. *Journal of the American Chemical Society*130, 4-5 (2007).

[38] Jha, A.K. & Prasad, K. Ferroelectric BaTiO₃ nanoparticles: Biosynthesis and characterization. *Colloids and Surfaces B: Biointerfaces*75, 330-334 (2010).

[39] Bansal, V., Poddar, P., Ahmad, A. & Sastry, M. Room-Temperature Biosynthesis of Ferroelectric Barium Titanate Nanoparticles. *Journal of the American Chemical Society*128, 11958-11963 (2006).

[40] Nuraje, N. et al. Room Temperature Synthesis of Ferroelectric Barium Titanate Nanoparticles Using Peptide Nanorings as Templates. *Advanced Materials*18, 807-811 (2006).

[41] Jha, A.K. & Prasad, K. Synthesis of BaTiO₃ nanoparticles: A new sustainable green approach. *Integrated Ferroelectrics*117, 49-54 (2010).

[42] Ould-Ely, T. et al. Large-scale engineered synthesis of BaTiO₃ nanoparticles using low-temperature bioinspired principles. *Nature Protocols*6, 97-104 (2011).

[43] Bansal, V., Bharde, A., Ramanathan, R. & Bhargava, S.K. Inorganic materials using 'unusual' microorganisms. *Advances in Colloid and Interface Science*.

[44] Kamiya, H. et al. Preparation of Highly Dispersed Ultrafine Barium Titanate Powder by Using Microbial-Derived Surfactant. *Journal of the American Ceramic Society*86, 2011-2018 (2012).

[45] Pithan, C., Hennings, D. & Waser, R. Progress in the Synthesis of Nanocrystalline BaTiO₃ Powders for MLCC. *International Journal of Applied Ceramic Technology*2, 1-14 (2005).

[46] Kim, P. et al. High Energy Density Nanocomposites Based on Surface-Modified BaTiO₃ and a Ferroelectric Polymer. *ACS Nano*3, 2581-2592 (2009).

[47] Dang, Z.-M. et al. Fabrication and Dielectric Characterization of Advanced BaTiO₃/Polyimide Nanocomposite Films with High Thermal Stability. *Advanced Functional Materials*18, 1509-1517 (2008).

[48] Zhou, T. et al. Improving Dielectric Properties of BaTiO₃/Ferroelectric Polymer Composites by Employing Surface Hydroxylated BaTiO₃ Nanoparticles. *ACS Applied Materials & Interfaces*3, 2184-2188 (2011).

[49] Fan, B.-H., Zha, J.-W., Wang, D.-R., Zhao, J. & Dang, Z.-M. Experimental study and theoretical prediction of dielectric permittivity in BaTiO₃/polyimide nanocomposite films. *Applied Physics Letters*100, 092903-4 (2012).

[50] Nagao, D., Kinoshita, T., Watanabe, A. & Konno, M. Fabrication of highly refractive, transparent BaTiO₃/poly(methyl methacrylate) composite films with high permittivities. *Polymer International*60, 1180-1184 (2011).

[51] Abe, K., Nagao, D., Watanabe, A. & Konno, M. Fabrication of highly refractive barium-titanate-incorporated polyimide nanocomposite films with high permittivity and thermal stability. *Polymer International*, n/a-n/a (2012).

[52] Balzani, V. Nanochemistry: A Chemical Approach to Nanomaterials. Geoffrey A. Ozin and André C. Arsenault. *Small*2, 678-679 (2006).

[53] Reznikov, Y. et al. Ferroelectric nematic suspension. *Applied Physics Letters*82, 1917-1919 (2003).

[54] Reznikov, Y. in Liquid Crystals Beyond Displays: Chemistry, Physics, and Applications (ed. Li, Q.) 403-426 (John Wiley & Sons, Inc., 2012).

[55] Liang, H.-H. & Lee, J.-Y. Enhanced Electro-Optical Properties of Liquid Crystals Devices by Doping with Ferroelectric Nanoparticles. *Ferroelectrics - Material Aspects*, 193-210 (2011).

[56] Ciofani, G. et al. Preparation of stable dispersion of barium titanate nanoparticles: Potential applications in biomedicine. *Colloids and Surfaces B: Biointerfaces*76, 535–543 (2010).

[57] Hsieh, C.-L., Grange, R., Pu, Y. & Psaltis, D. Bioconjugation of barium titanate nanocrystals with immunoglobulin G antibody for second harmonic radiation imaging probes. *Biomaterials*31, 2272–2277 (2010).

[58] Hsieh, C.-L., Grange, R., Pu, Y. & Psaltis, D. (eds. Campagnola, P.J., Stelzer, E.H.K. & von Bally, G.) 73670D-6 (SPIE, Munich, Germany, 2009).

[59] Hsieh, C.-L., Grange, R., Pu, Y. & Psaltis, D. Three-dimensional harmonic holographic microcopy using nanoparticles as probes for cell imaging. *Optics Express*17, 2880-2891 (2009).

[60] Grange, R., Hsieh, C.L., Pu, Y. & Psaltis, D. in European Conference on Lasers and Electro-Optics 2009 and the European Quantum Electronics Conference. CLEO Europe - EQEC 2009. 1-1 (2009).

[61] Ciofani, G. et al. in Piezoelectric Nanomaterials for Biomedical Applications 213-238 (Springer Berlin Heidelberg, 2012).

[62] Yust, B.G., Razavi, N., Pedraza, F. & Sardar, D.K. (eds. Cartwright, A.N. & Nicolau, D.V.) 82310H-6 (SPIE, San Francisco, California, USA, 2012).

[63] Zipfel, W.R. et al. Live tissue intrinsic emission microscopy using multiphoton-excited native fluorescence and second harmonic generation. *Proceedings of the National Academy of Sciences*100, 7075-7080 (2003).

[64] Pantazis, P., Maloney, J., Wu, D. & Fraser, S.E. Second harmonic generating (SHG) nanoprobes for in vivo imaging. *Proceedings of the National Academy of Sciences*107, 14535-14540 (2010).

[65] Ciofani, G., Ricotti, L. & Mattoli, V. Preparation, characterization and in vitro testing of poly(lactic-co-glycolic) acid/barium titanate nanoparticle composites for enhanced cellular proliferation. *Biomedical Microdevices*13, 255-266 (2011).

[66] Pantazis, P., Pu, Y., Psaltis, D. & Fraser, S. (eds. Periasamy, A. & So, P.T.C.) 71831P-5 (SPIE, San Jose, CA, USA, 2009).

[67] Horiuchi, N. Imaging: Second-harmonic nanoprobes. *Nature Photonics*5, 7-7 (2011).

[68] Hsieh, C.-L., Grange, R., Pu, Y. & Psaltis, D. (eds. Mohseni, H. & Razeghi, M.) 77590T-6 (SPIE, San Diego, California, USA, 2010).

[69] Hsieh, C.-L., Lanvin, T., Grange, R., Pu, Y. & Psaltis, D. in Conference on Lasers and Electro-Optics (CLEO), 2011 1-2 (2011).

[70] Semenov, S., Pham, N. & Egot-Lemaire, S. in World Congress on Medical Physics and Biomedical Engineering (eds. Dössel, O. & Schlegel, W.C.) 311-313 (Springer Berlin Heidelberg, Munich, Germany, 2009).

[71] Adur, J. et al. (eds. Periasamy, A., Konig, K. & So, P.T.C.) 82263P-7 (SPIE, San Francisco, California, USA, 2012).

[72] Wang, C.-C. et al. Differentiation of normal and cancerous lung tissues by multiphoton imaging. *Journal of Biomedical Optics*14, 044034-4 (2009).

[73] Chan, W.C.W. in Advances in Experimental Medicine and Biology 208 (Springer Science+Business Media, LLC, 2007).

[74] Dempsey, W.P., Fraser, S.E. & Pantazis, P. SHG nanoprobes: Advancing harmonic imaging in biology. *BioEssays*34, 351-360 (2012).

[75] Yust, B.G., Sardar, D.K. & Tsin, A. (eds. Cartwright, A.N. & Nicolau, D.V.) 79080G-7 (SPIE, San Francisco, California, USA, 2011).

[76] Ganeev, R.A., Suzuki, M., Baba, M., Ichihara, M. & Kuroda, H. Low- and high-order nonlinear optical properties of $BaTiO_3$ and $SrTiO_3$ nanoparticles. *Journal of the Optical Society of America B*25, 325-333 (2008).

[77] Lai, K.T., Nair, B.G. & Semenov, S. Optical and microwave studies of ferroelectric nanoparticles for application in biomedical imaging. *Microwave and Optical Technology Letters*54, 11-13 (2012).

[78] Ciofani, G. et al. 987-990 (KINTEX, Korea, 2010).

[79] Furuya, K., Morita, Y., Tanaka, K., Katayama, T. & Nakamachi, E. (eds. Martin-Palma, R.J. & Lakhtakia, A.) 79750U-6 (SPIE, San Diego, California, USA, 2011).

[80] Florence, A.T. "Targeting" nanoparticles: The constraints of physical laws and physical barriers. (2012).

Chapter 10

Photorefractive Effect in Ferroelectric Liquid Crystals

Takeo Sasaki

Additional information is available at the end of the chapter

1. Introduction

Ferroelectric liquid crystals have been attracting great interest for their application in photorefractive devices. The photorefractive effect is a phenomenon by which a change in refractive index is induced by the interference of laser beams [1, 2]. Dynamic holograms are easily realized by the photorefractive effect. Holograms generate three-dimensional images of objects. They are produced by recording interference fringes generated by light reflected from an object and a reference light (Figure 1). A hologram diffracts incident light to produce a three-dimensional images of an object. 3D displays are expected to be widely used as next-generation displays. However, current 3D displays are essentially stereograms. However, holographic displays that can realize natural 3D images are anticipated.

The photorefractive effect has the potential to realize dynamic holograms by recording holograms as a change in the refractive index of the medium [1, 2]. The photorefractive effect induces a change in the refractive index by a mechanism involving both photovoltaic and electro-optic effects (Figure 2). When two laser beams interfere in an organic photorefractive material, a charge generation occurs at the bright positions of the interference fringes. The generated charges diffuse or drift within the material. Since the mobilities of positive and negative charges are different in most organic materials, a charge separated state is formed. The charge with the higher mobility diffuses over a longer distance than the charge with lower mobility, such that while the low mobility charge stays in the bright areas, the high mobility charge moves to the dark areas. The bright and dark positions of the interference fringes are thus charged with opposite polarities and an internal electric field (space-charge field) is generated in the area between the bright and dark positions. The refractive index of this area between the bright and dark positions is changed through the electro-optic effect. Thus, a refractive index grating (or hologram) is formed. One material class that exhibits high photorefractivity is that of glassy photoconductive polymers doped with high concentrations of D-π-A chromophores, in which donor and acceptor groups are attached to a πconjugate system [3-5]. In order to obtain photorefractivity in polymer materials, a high electric field of 10–50 V/μm is usually applied to a polymer film [6-8]. This electric field is necessary to increase charge generation efficiency. Significantly, the photorefractive effect permits two-beam coupling. It can be used to coherently amplify signal beams, and so has the potential to be used in a wide range of optical technologies as a transistor in electric circuits. With the aim of developing a 3D display, a multiplex hologram has been demonstrated using a photorefractive polymer film [9,10]. Clear 3D images were recorded in the film. However, the slow response (~100 ms) and the high electric field (30–50 V/μm) required to activate the photorefractive effect in the polymer materials both need to be improved.

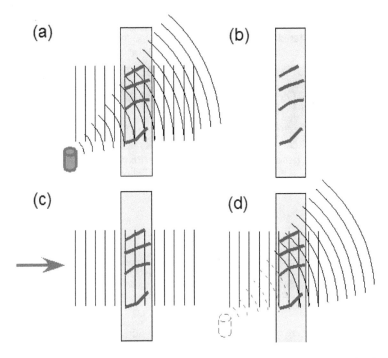

Figure 1. Hologram recording and image reconstruction. (a) A light reflected from an object is interfered with by a reference light in a photo-sensitive material, such as a photopolymer. (b) The interference fringe is recorded in a photosensitive material (hologram). (c) A readout beam is irradiated by the hologram. (d) The readout beam is diffracted by the hologram and the object image is reconstructed.

Ferroelectric liquid crystals are expected to be used as high performance photorefractive materials [11-13]. A photorefractive ferroelectric liquid crystal with a fast response of 5 ms has been reported [14]. The photorefractive effect has been reported in surface-stabilized ferroelectric liquid crystals (SS-FLCs) doped with a photoconductive compound. Liquid crystals are classified into several groups. The most well known are nematic liquid crystals and smectic liquid crystals.

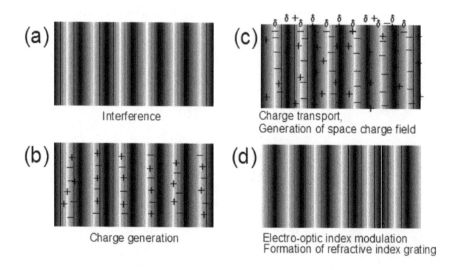

Figure 2. Schematic illustration of the mechanism of the photorefractive effect. (a) Two laser beams interfere in the photorefractive material; (b) charge generation occurs at the light areas of the interference fringes; (c) electrons are trapped at the trap sites in the light areas and holes migrate by diffusion or drift in the presence of an external electric field and generate an internal electric field between the light and dark positions; (d) the refractive index of the corresponding area is altered by the internal electric field generated.

Nematic liquid crystals are used in LC displays. On the other hand, smectic liquid crystals are very viscous and, hence, are not utilized in any practical applications. Ferroelectric liquid crystals (FLCs) belong to the class of smectic liquid crystals that have a layered structure (Figure 3). The molecular structure of a typical FLC contains a chiral unit, a carbonyl group, a central core - which is a rigid rod-like structure, such as biphenyl, phenylpyrimidine or phenylbenzoate - and a flexible alkyl chain [15]. Thus, the dipole moment of an FLC molecule is perpendicular to the molecular long axis. FLCs exhibit a chiral smectic C phase (SmC*) that possesses a helical structure. Compared to nematic LCs, FLCs are more crystal than liquid and the preparation of fine FLC films requires several sophisticated techniques. Obtaining a uniformly aligned, defect-free, surface-stabilized FLC (SS-FLC, Figure 3) using a single FLC compound is very difficult, and mixtures of several LC compounds are usually used to obtain fine SS-FLC films.

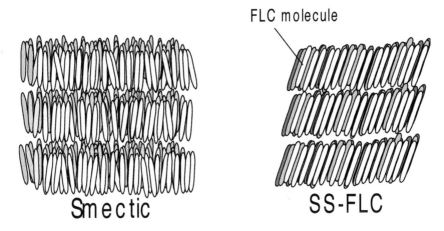

Figure 3. Structures of the smectic phase and the surface stabilized smectic-C phase (SS-FLC).

The FLC mixtures are composed of the base LC - which is also a mixture of several LC compounds - and a chiral compound. The chiral compound introduces a helical structure into the LC phase through supramolecular interactions. It should be mentioned here that in order to observe ferroelectricity in these materials, the ferroelectric liquid crystals must be formed into thin films. The thickness of the film must be within a few micrometers. When an FLC is sandwiched between glass plates to form a film a few micrometers thick, the helical structure of the smectic C phase uncoils and a surface-stabilized state (SS-state) is formed in which spontaneous polarization (Ps) appears. For display applications, the thickness of the film is usually 2 μm. In such thin films, FLC molecules can align in only two directions. This state is called a surface-stabilized state (SS-state). The alignment direction of the FLC molecules changes according to the direction of the spontaneous polarization (Figure 4). The direction of the spontaneous polarization is governed by the photoinduced internal electric field, giving rise to a refractive index grating with properties dependent on the direction of polarization.

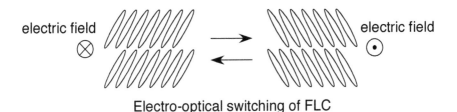

Figure 4. Electro-optical switching in the surface-stabilized state of FLCs.

Figure 5 shows a schematic illustration of the mechanism of the photorefractive effect in FLCs. When laser beams interfere in a mixture of an FLC and a photoconductive compound, charge separation occurs between bright and dark positions and an internal electric field is produced. The internal electric field alters the direction of spontaneous polarization in the area between the bright and dark positions of the interference fringes, which induces a periodic change in the orientation of the FLC molecules. This is different from the processes that occur in other photorefractive materials in that the molecular dipole rather than the bulk polarization responds to the internal electric field. Since the switching of FLC molecules is due to the response of bulk polarization, the switching is extremely fast.

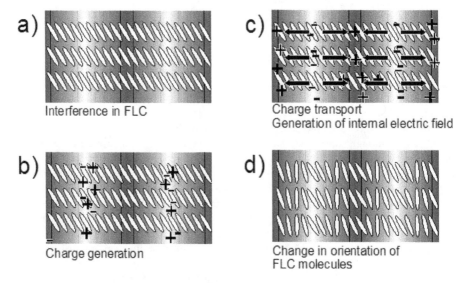

a)

Interference in FLC

c)

Charge transport
Generation of internal electric field

b)

Charge generation

d)

Change in orientation of
FLC molecules

Figure 5. Schematic illustration of the mechanism of the photorefractive effect in FLCs. (a) Two laser beams interfere in the surface-stabilized state of the FLC/photoconductive compound mixture; (b) charge generation occurs at the bright areas of the interference fringes; (c) electrons are trapped at the trap sites in the bright areas, holes migrate by diffusion or drift in the presence of an external electric field to generate an internal electric field between the bright and dark positions; (d) the orientation of the spontaneous polarization vector (i.e., the orientation of mesogens in the FLCs) is altered by the internal electric field.

2. Characteristics of the photorefractive effect

Since a change in the refractive index via the photorefractive effect occurs in the areas between the bright and dark positions of the interference fringe, the phase of the resulting index grating is shifted from the interference fringe. This is characteristic of the photorefractive effect that the phase of the refractive index grating is $\pi/2$-shifted from the interference fringe. When the material is photochemically active and is not photorefractive, a photochemical reaction takes place at the bright areas and a refractive index grating with the same phase as that of the interference fringe is formed (Figure 6(a)).

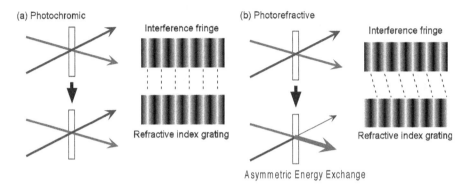

Figure 6. (a) Photochromic grating, and (b) photorefractive grating.

The interfering laser beams are diffracted by this grating; however, the apparent transmitted intensities of the laser beams do not change because the diffraction is symmetric. Beam 1 is diffracted in the direction of beam 2 and beam 2 is diffracted in the direction of beam 1. However, if the material is photorefractive, the phase of the refractive index grating is shifted from that of the interference fringes, and this affects the propagation of the two beams. Beam 1 is energetically coupled with beam 2 for the two laser beams. Consequently, the apparent transmitted intensity of beam 1 increases and that of beam 2 decreases (Figure 6(b)). This phenomenon is termed 'asymmetric energy exchange' in the two-beam coupling experiment. The photorefractivity of a material is confirmed by the occurrence of this asymmetric energy exchange.

3. Measurement of photorefractivity

The photorefractive effect is evaluated by a two-beam coupling method and by a four-wave mixing experiment. Figure 7(a) shows a schematic illustration of the experimental setup used for the two-beam coupling method. A p-polarized beam from a laser is divided into two beams by a beam splitter and the beams are interfered within the sample film. An electric field is applied to the sample using a high voltage supply unit. This external electric field is applied in order to increase the efficiency of charge generation in the film. The change in the transmitted beam intensity is monitored. If a material is photorefractive, an asymmetric energy exchange is observed. The magnitude of photorefractivity is evaluated using a parameter called the gain coefficient, which is calculated from the change in the transmitted intensity of the laser beams induced through the two-beam coupling [1]. In order to calculate the two-beam coupling gain coefficient, it must be determined whether the diffraction condition is within the Bragg regime or within the Raman-Nath regime. These diffraction conditions are distinguished by a dimensionless parameter Q.

$$Q = 2\pi l L / n \Lambda^2 \qquad (1)$$

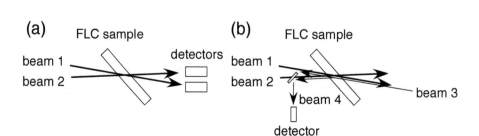

Figure 7. Schematic illustrations of the experimental set-up for the (a) two-beam coupling, and (b) four-wave-mixing techniques.

Q > 1 is defined as the Bragg regime of optical diffraction. In this regime, multiple scattering is not permitted, and only one order of diffraction is produced. Conversely, Q < 1 is defined as the Raman-Nath regime of optical diffraction. In this regime, many orders of diffraction can be observed. Usually, Q > 10 is required to guarantee that the diffraction is entirely within the Bragg regime. When the diffraction is in the Bragg diffraction regime, the two-beam coupling gain coefficient Γ (cm⁻¹) is calculated according to the following equation:

$$\Gamma = \frac{1}{D} \ln \left(\frac{gm}{1 + m - g} \right) \qquad (2)$$

where D = L/cos(θ) is the interaction path for the signal beam (L=sample thickness, θ =propagation angle of the signal beam in the sample), g is the ratio of the intensities of the signal beam behind the sample with and without a pump beam, and m is the ratio of the beam intensities (pump/signal) in front of the sample.

A schematic illustration of the experimental setup used for the four-wave mixing experiment is shown in Figure 7(b). S-polarized writing beams are interfered in the sample film and the diffraction of a p-polarized probe beam, counter-propagating to one of the writing beams, is measured. The diffracted beam intensity is typically measured as a function of time, applied (external) electric field and writing beam intensities, etc. The diffraction efficiency is defined as the ratio of the intensity of the diffracted beam and the intensity of the probe beam that is transmitted when no grating is present in the sample due to the writing beams. In probing the grating, it is important that beam 3 does not affect the grating or interact with the writing beams. This can be ensured by making the probe beam much weaker than the writing beams and by having the probe beam polarized orthogonal to the writing beams.

4. Photorefractive effect of FLCs

4.1. Two-beam coupling experiments on FLCs

The photorefractive effect in an FLC was first reported by Wasielewsky et al. in 2000 [11]. Since then, details of photorefractivity in FLC materials have been further investigated by Sasaki et al. and Golemme et al. [12-14]. The photorefractive effect in a mixture of an FLC and a photoconductive compound was measured in a two-beam coupling experiment using a 488 nm Ar⁺ laser. The structures of the photoconductive compounds used are shown in Figure 8. A commercially available FLC, SCE8 (Clariant), was used. CDH was used as a photoconductive compound and TNF was used as a sensitizer. The concentrations of CDH and TNF were 2 wt% and 0.1 wt% respectively. The samples were injected into a 10-μm-gap glass cell equipped with 1 cm² ITO electrodes and a polyimide alignment layer (Figure 9).

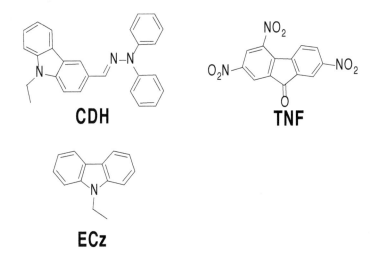

Figure 8. Structures of the photoconductive compound CDH, ECz and the sensitizer TNF

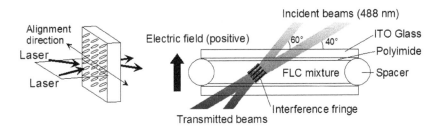

Figure 9. Laser beam incidence condition and the structure of the LC cell.

Figure 10 shows a typical example of asymmetric energy exchange observed in the
FLC(SCE8)/CDH/TNF sample under an applied DC electric field of 0.1 V/μm [12]. The interference of the divided beams in the sample resulted in the increased transmittance of one
beam and the decreased transmittance of the other. The change in the transmitted intensities
of the two beams is completely symmetric, as can be seen in Figure 10. This indicates that
the phase of the refractive index grating is shifted from that of the interference fringes. The
grating formation was within the Bragg diffraction regime and no higher-order diffraction
was observed under the conditions used.

The temperature dependence of the gain coefficient of SCE8 doped with 2 wt% CDH and 0.1
wt% TNF is shown in Figure 11(a). Asymmetric energy exchange was observed only at temperatures below 46°C. The spontaneous polarization of the identical sample is plotted as a
function of temperature in Figure 11(b).

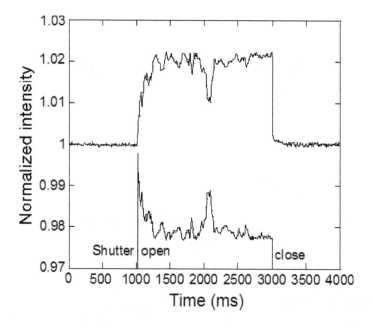

Figure 10. A typical example of asymmetric energy exchange observed in an FLC (SCE8) mixed with 2 wt% CDH and
0.1 wt% TNF. An electric field of +0.3 V/μm was applied to the sample.

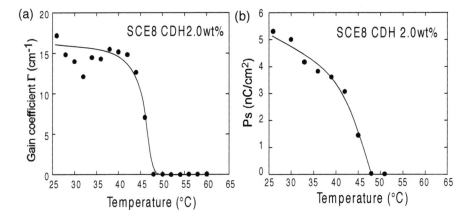

Figure 11. Temperature dependence of (a) the gain coefficient, and (b) the spontaneous polarization of an FLC (SCE8) mixed with 2 wt% CDH and 0.1 wt% TNF. For two-beam coupling experiments, an electric field of 0.1 V/μm was applied to the sample.

Similarly, the spontaneous polarization vanished when the temperature was raised above 46°C. Thus, asymmetric energy exchange was observed only in the temperature range in which the sample exhibited ferroelectric properties; in other words, the SmC* phase. Since the molecular dipole moment of the FLCs is small and the dipole moment is aligned perpendicular to the molecular axis, large changes in the orientation of the molecular axis cannot be induced by an internal electric field in the SmA or N* phase of the FLCs. However, in the SmC* phase, reorientation associated with spontaneous polarization occurs due to the internal electric field. The spontaneous polarization also causes the orientation of FLC molecules in the corresponding area to change accordingly. A maximum resolution of 0.8 μm was obtained in this sample.

4.2. Effect of the magnitude of the applied electric field

In polymeric photorefractive materials, the strength of the externally applied electric field is a very important factor. The external electric field is necessary to increase the charge separation efficiency sufficiently to induce a photorefractive effect. In other words, the photorefractivity of the polymer is obtained only with the application of a few V/μm electric fields. The thickness of the polymeric photorefractive material commonly reported is about 100 μm, so the voltage necessary to induce the photorefractive effect is a few kV. On the other hand, the photorefractive effect in FLCs can be induced by applying a very weak external electric field. The maximum gain coefficient for the FLC (SCE8) sample was obtained using an electric field strength of only 0.2–0.4 V/μm. The thickness of the FLC sample is typically 10 μm, so that the voltage necessary to induce the photorefractive effect is only a few V. The dependence of the gain coefficient of a mixture of FLC (SCE8)/CDH/TNF on the strength of the electric field is shown in Figure 12. The gain coefficient of SCE8 doped with 0.5–1 wt% CDH increased with the strength of the external electric field. However, the gain coefficient of

SCE8 doped with 2 wt% CDH decreased when the external electric field exceeded 0.4 V/μm. The same tendency was observed for M4851/050 as well. The formation of an orientational grating is enhanced when the external electric field is increased from 0 to 0.2 V/μm as a result of induced charge separation under a higher external electric field. However, when the external electric field exceeded 0.2 V/μm, a number of zigzag defects appeared in the surface-stabilized state. These defects cause light scattering and result in a decrease in the gain coefficient.

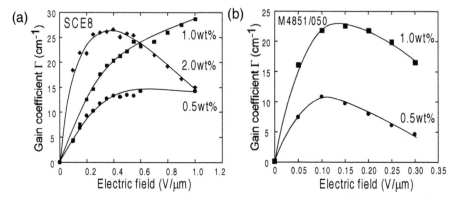

Figure 12. Electric field dependence of the gain coefficient of SCE8 and M4851/050 mixed with several concentrations of CDH and 0.1 wt% TNF in a 10 μm-gap cell measured at 30 °C.

4.3. Refractive index grating formation time

The formation of a refractive index grating involves charge separation and reorientation. The index grating formation time is affected by these two processes and both may be rate-determining steps. The refractive index grating formation times in SCE8 and M4851/050 were determined based on the simplest single-carrier model of photorefractivity [1,2], wherein the gain transient is exponential. The rising signal of the diffracted beam was fitted using a single exponential function, shown in equation (3).

$$\gamma(t) - 1 = (\gamma - 1)[1 - \exp(-t/\tau)]^2 \tag{3}$$

Here, $\gamma(t)$ represents the transmitted beam intensity at time t divided by the initial intensity ($\gamma(t) = I(t)/I0$) and τ is the formation time. The grating formation time in SCE8/CDH/TNF is plotted as a function of the strength of the external electric field in Figure 13(a). The grating formation time decreased with increasing electric field strength due to the increased efficiency of charge generation. The formation time was shorter at higher temperatures, corresponding to a decrease in the viscosity of the FLC with increasing temperature. The formation time for SCE8 was found to be 20 ms at 30°C. As shown in Figure 13(b), the formation time for M4851/050 was found to be independent of the magnitude of the external electric field, with a time of 80-90 ms for M4851/050 doped with 1 wt% CDH and 0.1 wt% TNF. This is slower than for SCE8, although the spontaneous polarization of M4851/050 (-14 nC/cm^2) is larger than that of SCE8 (-4.5 nC/cm^2) and the response time of the electro-optical switching (the flipping of spontaneous polarization) to an electric field (±10 V in a 2 μm cell) is shorter for M4851/050. The slower formation of the refractive index grating in M4851/050 is likely due to the poor homogeneity of the SS-state and charge mobility.

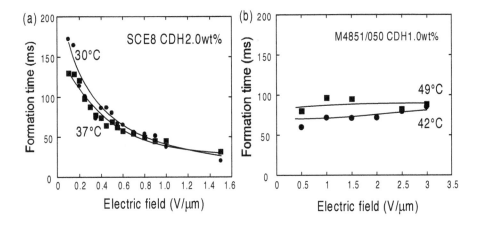

Figure 13. Electric field dependence of the index grating formation time. (a) SCE8 mixed with 2 wt% CDH and 0.1 wt % TNF in the two-beam coupling experiment. •, measured at 30 °C (T/T$_{SmC^*-SmA}$=0.95); ■, measured at 36 °C (T/T$_{SmC^*-SmA}$=0.97). (b) M4851/050 mixed with 1 wt% CDH and 0.1 wt% TNF in a two-beam coupling experiment. •, measured at 42 °C (T/T$_{SmC^*-SmA}$=0.95); ■, measured at 49 °C (T/T$_{SmC^*-SmA}$=0.97).

4.4. Formation mechanism of the internal electric field in FLCs

Since the photorefractive effect is induced by the photoinduced internal electric field, the mechanism of the formation of the space-charge field in the FLC medium is important. The two-beam coupling gain coefficients of mixtures of FLC (SCE8) and photoconductive compounds under a DC field were investigated as a function of the concentration of TNF (electron acceptor). The photoconductive compounds - CDH, ECz and TNF (Figure 8) - were used in this examination. When an electron donor with a large molecular size (CDH) relative to the TNF was used as the photoconductive compound, the gain coefficient was strongly affected by the concentration of TNF (Figure 14(a)). However, when ethylcarbazole (ECz) - the molecular size of which is almost the same as that of TNF - was used, the gain coefficient was less affected by the TNF concentration (Figure 14(b)).

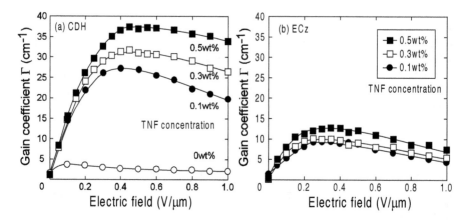

Figure 14. Dependence of the TNF concentration on the gain coefficients of an FLC doped with photoconductive dopants. (a) SCE8 doped with 2 wt% CDH, and (b) SCE8 doped with 2 wt% ECz. An electric field of ± 0.5 V/μm, 100 Hz was applied.

These findings suggest that ionic conduction plays a major role in the formation of the space-charge field. The mobility of the CDH cation is smaller than that of the TNF anion, and this difference in mobility is thought to be the origin of the charge separation. In this case, the magnitude of the internal electric field is dominated by the concentration of the ionic species. On the other hand, the difference in the mobilities of ECz and TNF is small and, thus, less effective charge separation is induced, indicating that the internal electric field is independent of the concentration of ionic species.

4.5. Photorefractive effect in FLC mixtures containing photoconductive chiral compounds

4.5.1. Photoconductive chiral dopants

The photorefractive effect of FLC mixtures containing photoconductive chiral dopants has been investigated [15]. The structures of the LC compounds, the electron acceptor trinitro-fluorenone (TNF) and the photoconductive chiral compounds are shown in Figure 15. The mixing ratio of 8PP8 and 8PP10 was set to 1:1 because the 1:1 mixture exhibits the SmC phase over the widest temperature range. Hereafter, the 1:1 mixture of 8PP8 and 8PP10 is referred to as the base LC. The concentration of TNF was 0.1 wt%. Four photoconductive chiral compounds with the terthiophene chromophore (3T-2MB, 3T-2OC, 3T-OXO, and 3T-CF3) were synthesized. The base LC, TNF and a photoconductive chiral compound were dissolved in dichloroethane and the solvent was evaporated. The mixture was then dried in a vacuum at room temperature for one week. The samples were subsequently injected into a 10-μm-gap glass cell equipped with 1-cm² ITO electrodes and a polyimide align-ment layer for the measurements. The base LC, which is a 1:1 mixture of 8PP8 and 8PP10, was mixed with the photoconductive chiral dopants and the electron acceptor (TNF). The concentration of TNF was set to 0.1 wt%. The terthiophene chiral dopants showed high miscibility with the phenyl pyrimidine-type smectic LC. The chiral smectic C (SmC*) phase appeared in all of the mixtures of the base LC and the chiral dopants. With the increase of the concentration of the chiral dopants, the temperature range of the SmC* phase and the chiral nematic (N*) phase were reduced, whereas that of the smectic A (SmA) phase was enhanced. The miscibility of the 3T-CF3 with the base LC was the lowest among the four chiral dopants. It was considered that the dipole moment of the trifluoromethyl substituted group is large, so that the 3T-CF3 molecules tend to aggregate. All the samples exhibited absorption maxima at 394 nm. Absorption at 488 nm (wavelength of the laser used) was small. The absorption spectra were not changed when TNF (1.0×10-5 mol/L) was added to the solution. The small absorption at the laser wavelength is advantageous for minimizing the optical loss. The photocurrents in the mixtures of the base LC, photoconductive chiral dopants and TNF were measured. As shown in Figure 16, the samples were good insulators in the dark. When a 488 nm laser irradiated the samples, photocurrents were clearly observed. The magnitudes of the photocurrents were slightly different in the four samples. The only difference in the molecular structures of these compounds is the chiral substituent. Thus, the difference in the photocurrent cannot be attributed to the difference in the molecular structure. It was considered that the miscibility of the photoconductive chiral compounds to the LC and the homogeneity of the LC phase affected the magnitude of the photocurrent.

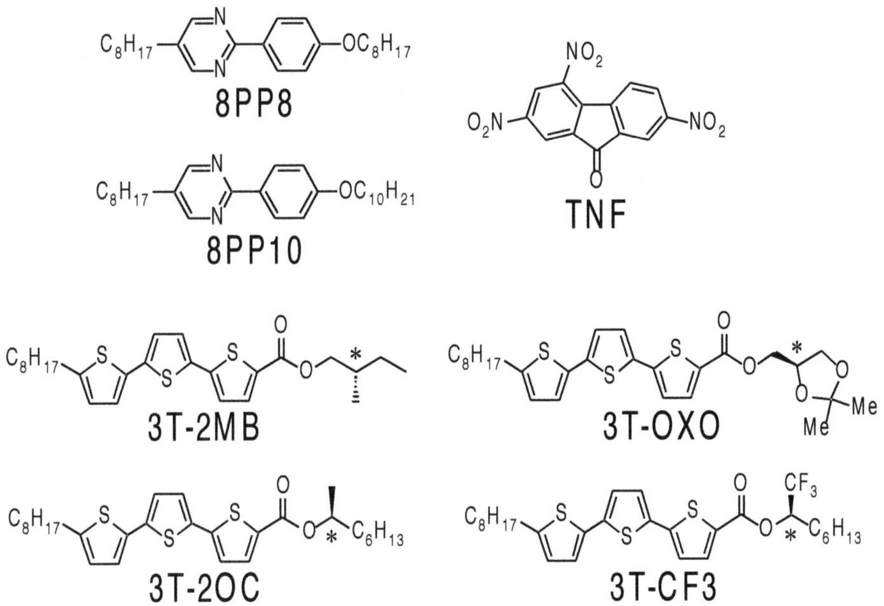

Figure 15. Structures of the smectic LCs (8PP8 and 8PP10), photoconductive chiral dopants (3T-2MB, 3T-2OC, 3T-OXO and 3T-CF3) and the sensitizer TNF used in this work.

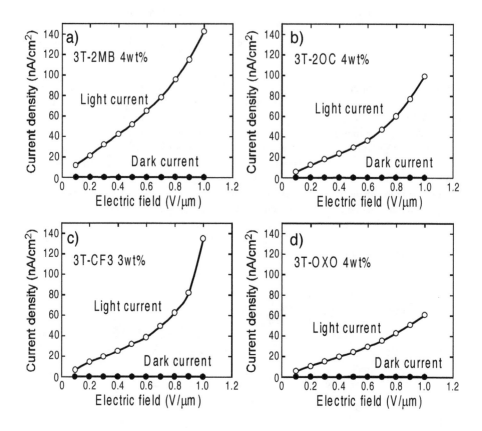

Figure 16. Magnitudes of light-current and dark-current of mixtures of the base LC, photoconductive chiral compound and TNF measured in a 10-μm-gap LC cell as a function of the external electric field. An electric field of 0.1 V/μm was applied. A 488 nm Ar⁺ laser (10 mW/cm², 1 mm diameter) was used as the irradiation source.

4.5.2. Two-beam coupling experiment on photoconductive FLC mixtures

The photorefractive effect was measured in a two-beam coupling experiment. A linearly polarized beam from an Ar⁺ laser (488 nm, continuous wave) was divided into two by a beam splitter, each of which was then interfered in the sample film. A p-polarized beam was used. The laser intensity was 2.5 mW for each beam (1 mm diameter). The incident beam angles to the glass plane were 30° and 50°. Each interval of the interference fringe was 1.87 μm. Figure 17 shows typical examples of the asymmetric energy exchange observed in a mixture of the base LC, 3T-2MB and TNF at 30 ºC with the application of an electric field of 0.2–0.4 V/μm.

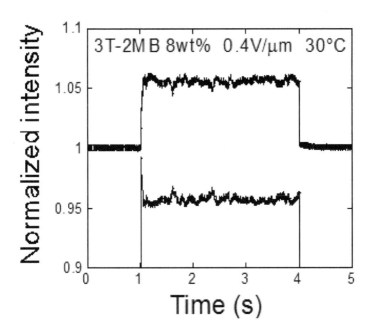

Figure 17. Examples of the results of two-beam coupling experiments for mixtures of the base LC, 3T-2MB and TNF measured at 30 °C.

The interference of the divided beams in the sample resulted in the increased transmittance of one of the beams and the decreased transmittance of the other beam. These transmittance characteristics were reversed when the polarity of the applied electric field was reversed. Asymmetric energy exchange was only observed when an electric field was applied, indicating that beam coupling was not caused by a thermal grating. With an increased concentration of 3T-2MB, the magnitude of the gain coefficient also increased. In order to calculate the two-beam coupling gain coefficient, the diffraction condition needs to be correctly identified.

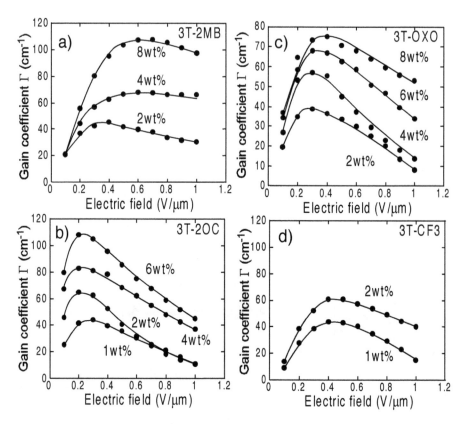

Figure 18. Electric field dependence of the gain coefficients of the mixtures of the base LC, photoconductive chiral compounds and TNF. (a) 3T-2MB; (b) 3T-2OC; (c) 3T-OXO; (d) 3T-CF3.

The difference in the gain coefficients in mixtures of the base LC, photoconductive chiral dopants (3T-2MB, 3T-2OC, 3T-OXO, and 3T-CF3) and TNF was investigated. All the samples formed a finely aligned SS-state in 10-μm-gap cells with a LX-1400 polyimide alignment layer and exhibited clear photorefractivity in the ferroelectric phase. The asymmetric energy exchange was only observed in the temperature range in which the sample exhibits ferroelectric properties (SmC* phase). The gain coefficients of the samples are plotted as a function of the magnitude of the external electric field in Figure 18. The gain coefficient increased with the strength of the external electric field up to 0.2-0.6 V/μm and then decreased with the strength of the external electric field. As the concentration of the photoconductive chiral dopants increased, so did the gain coefficient. This may be due to the increased density of charge carriers in the FLC medium and an increase in the magnitude of Ps. All the samples exhibited relatively large photorefractivity. A gain coefficient higher than 100 cm^{-1} was obtained in the 3T-2OC (6 wt%) sample with the application of only 0.2 V/μm. This means that a voltage of only 2 V is needed to obtain the gain coefficient of 100 cm^{-1} in a 10 μm FLC sample. In the FLCs reported previously, gain coefficients of only 50-60 cm^{-1} were obtained with the application of a 1 V/μm electric field. A gain coefficient higher than 100 cm^{-1} was also obtained in the 3T-2MB sample with an applied electric field of 0.5 V/μm. In order to obtain photorefractivity in polymer materials, the application of a high electric field of 10-50 V/μm to the polymer film is typically required. The small electric field necessary for the activation of the photorefractive effect in FLCs is thus a great advantage for their use in photorefractive devices. The miscibility of 3T-CF3 with the base LC was low and could be mixed with the base LC at concentrations lower than 2 wt%. The grating formation time in the mixtures of the base LC, photoconductive chiral dopants and TNF is plotted as a function of the strength of the external electric field in Figure 19. The grating formation time decreased with an increased electric field strength due to the increased efficiency of charge generation. The shortest formation time was obtained as 5–8 ms with a 1 V/μm external electric field in all chiral compounds. The 3T-CF3 sample exhibited the fastest response. This was because of the larger polarity of the 3T-CF3.

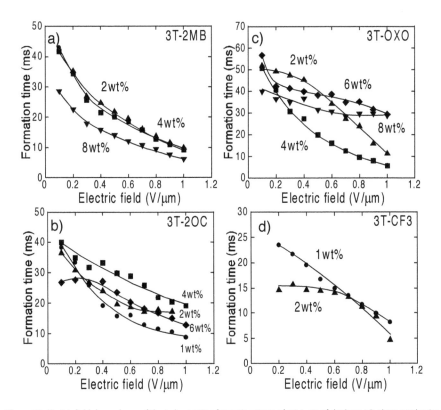

Figure 19. Electric field dependence of the index grating formation times of mixtures of the base LC, photoconductive chiral compounds and TNF measured at 30 °C. (a) 3T-2MB; (b) 3T-2OC; (c) 3T-OXO; (d) 3T-CF3.

4.5.3. Photorefractive effect in a ternary mixture of a SmC liquid crystal doped with the photoconductive chiral compound

The photorefractive effect of the mixture shown in Figure 20 was investigated. The gain co-efficients of the samples were measured as a function of the applied electric field strength (Figure 21(a)). The gain coefficient was calculated to be higher than 800 cm^{-1} in the 10 wt% sample with the application of only 1 V/μm. This gain coefficient is eight times higher than that of the FLCs described in the previous section. It was considered that a higher transparency of the ternary LC mixture contributed to the high gain coefficient. The small electric field required to activate the photorefractive effect in FLCs is a great advantage for photorefractive devices. The response time decreased with an increased electric field strength due to the increased charge separation efficiency (Figure 21(b)). The shortest formation time obtained was 8 ms for an external electric field of 1.5 V/μm. The large gain and fast response are advantageous for realizing optical devices, such as real-time image amplifiers and accurate measurement devices.

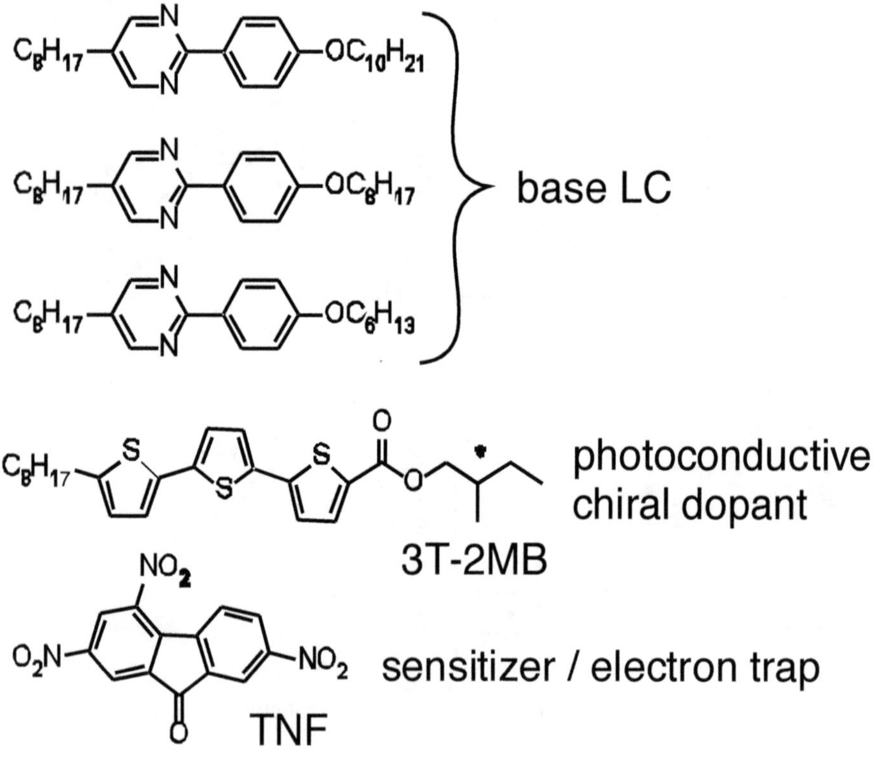

Figure 20. Photorefractive FLC mixture containing a ternary mixture of smectic LCs.

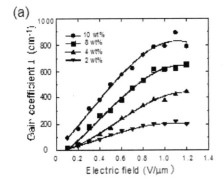 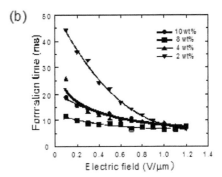

Figure 21. (a) Electric field dependence of gain coefficients of mixtures of the base LC, 3T-2MB and TNF (0.1 wt%) measured at 30 °C. The 3T-2MB concentration was within the range 2–10 wt%. (b) Refractive index grating formation time (response time) of mixtures of the base LC, 3T-2MB and TNF (0.1 wt%) measured at 30 °C. The 3T-2MB concentration was within the range 2–10 wt%.

4.5.4. Formation of dynamic holograms in FLC mixtures

The formation of a dynamic hologram was demonstrated. A computer-generated animation was displayed on a spatial light modulator (SLM). A 488 nm beam from a DPSS laser was irradiated on the SLM and the reflected beam was incident on the FLC sample. A reference beam interfered with the beam from the SLM in the FLC sample. The refractive index grating formed was within the Raman–Nath regime, in which multiple scattering is allowed. A 633 nm beam from a He–Ne laser was irradiated on the FLC sample and the diffraction was observed. A moving image of the animation on the computer monitor was observed in the diffracted beam (Figure 22). No image retention was observed, which means that the hologram image (refractive index grating) formed in the FLC was rewritten sufficiently rapidly to project the reproduction of a smooth holographic movie.

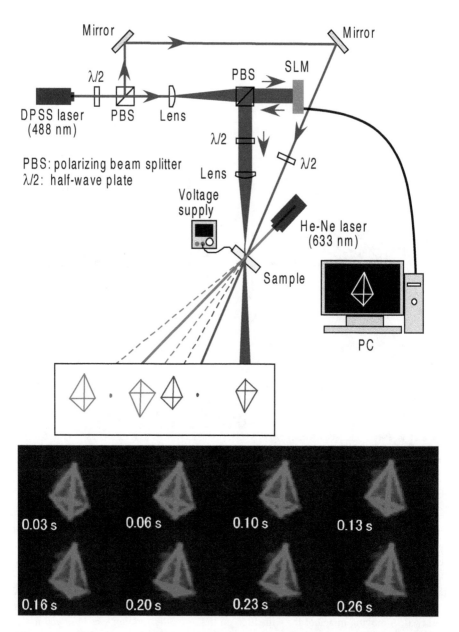

Figure 22. Dynamic hologram formation experiment on an FLC sample. A computer-generated animation was displayed on the SLM. The SLM modulated the object beam (488 nm), which was irradiated on the FLC sample and interfered with the reference beam. The readout beam (633 nm) was irradiated on the FLC and diffraction was observed.

5. Photorefractive effect in FLCs with the application of an AC field

5.1. Formation of dynamic holograms based on the spatial modulation of the molecular motion of FLCs

The formation of dynamic holograms based on the spatial modulation of the molecular motion of ferroelectric liquid crystals (FLCs) was demonstrated [16, 17]. The switching movement of an FLC molecule is essentially a rotational motion along a conical surface. When an alternating triangular-waveform voltage is applied, the FLC molecules uniformly go through a consecutive rotational switching motion along a conical surface (Figure 23(a)). If a photoconductive FLC material is used, this rotational motion can be modulated by illuminating the material with interfering laser beams, as shown in Figure 23. The internal electric field vector is directed along the interference fringe wave vector and, in many cases, it differs from the direction of the applied alternating electric field. Thus, the total electric field at each moment on the FLC molecules is altered by the presence of the internal electric field. The consecutive rotational motion of the FLC molecules in the areas between the light and dark positions of the interference fringes is biased by the internal electric field. Consequently, a grating based on the spatial difference in the rotational motion (or switching motion) of the FLC molecules is created. This grating is different from those currently used for holograms, wherein changes in the static properties of a medium - such as absorbance, transparency, film thickness and molecular orientation - are induced by photochemical reactions.

The formation of a motion-mode grating was examined using a SCE8/CDH/TNF mixture in a 10 μm-gap-cell. The formation of a holographic grating and the occurrence of a phase shift between the formed grating and the interference fringes were examined in this experiment. An alternating triangular-waveform electric field (0 to ±1 V/μm, and 1 kHz to 3 MHz) was applied to the sample. Under the effect of an alternating triangular-waveform electric field of ±0.5 V/μm - 100 kHz in the two-beam coupling experiment - the FLC molecules exhibited a consecutive switching motion. This switching was confirmed using a polarizing microscope equipped with a photodetector Figure 24. Figure 25(a) shows the transmitted intensities of the laser beams through the FLC/CDH/TNF mixture upon the application of an alternating electric field as a function of time. The interference of the divided beams in the sample resulted in the increased transmittance of one of the beams and the decreased transmittance of the other. The incident beam conditions were the same as those used in the DC experiment.

Although the transmitted intensity of the laser beam oscillates due to the switching motion of the FLC molecules, the average intensities of the beams were symmetrically changed, as shown in Figure 25. This indicates that a diffraction grating based on the spatial difference in the rotational switching motion of FLC molecules was formed. The symmetric change in the transmittance of the two beams proves that the phase of the motion-mode grating shifted from that of the interference fringe. No higher-order diffraction was observed under this condition. The spacing of the grating was calculated to be 0.9 μm (1,100 lines/mm). Figure 26 shows the temperature dependence of the gain coefficient obtained under an AC electric field. Asymmetric energy exchange was observed only at those temperatures at which the FLC/CDH/TNF mixture exhibits a ferroelectric phase. This suggests that ferroelectricity or the switching movement of SmC* is necessary for beam coupling under an AC electric field.

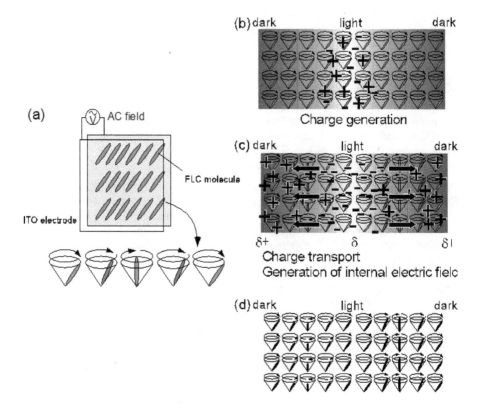

Figure 23. (a) Electro-optical switching of an FLC. (b) The rotational motion of FLC molecules under the application of an alternating electric field. (c) Positive and negative charges appear at the light positions of the interference fringe. (d) An internal electric field develops in the area between the light and dark positions of the interference fringes. The rotational motion of the FLC molecules in the corresponding area is biased by the internal electric field.

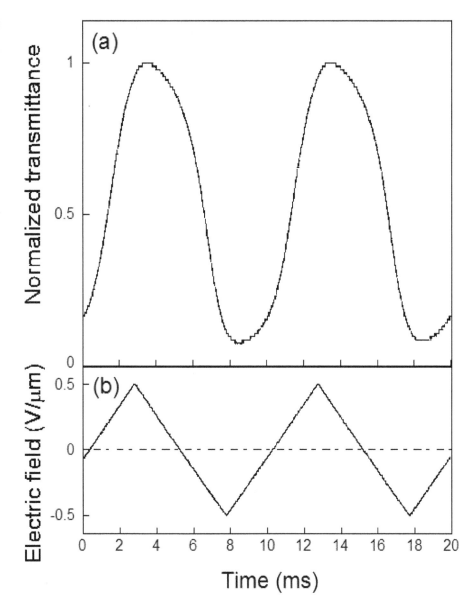

Figure 24. Switching behaviour of SCE8 mixed with 2 wt% CDH and 0.1 wt% TNF under a 100 Hz, ± 0.5 Vm/μm triangular wave electric field. (a) Transmittance of light through the FLC under a polarizing microscope. (b) Magnitude of the applied electric field.

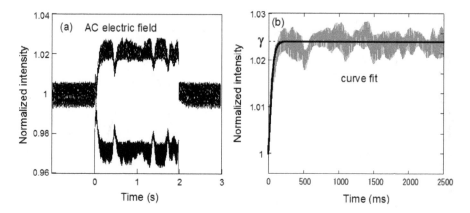

Figure 25. a) Typical example of the asymmetric energy exchange observed in two-beam coupling experiments. A triangular waveform AC electric field of ± 0.5 Vm/μm, 100 Hz was applied in this case. The sample angle α was 50° and the intersection angle φ was 20°. The shutter was opened at t = 0 s and closed at t = 2 s. (b) An example of the curve fit.

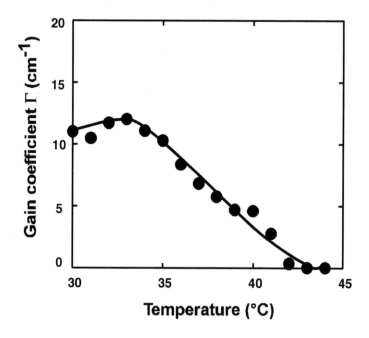

Figure 26. Gain coefficient of the two-beam coupling under an AC electric field as a function of temperature. An AC electric field of ± 0.5 V/μm, 100 Hz was applied.

Figure 27 shows the gain coefficient plotted as a function of the electric field strength. Asymmetric energy exchange was observed at electric field strengths higher than ±0.1 V/μm. Furthermore, no asymmetric energy exchange was observed without an external electric field. This eliminates the possibility that beam coupling resulted from either a thermal grating or from a photochemically-formed grating. The gain coefficient was independent of the external electric field strength for fields higher than ±0.1 V/μm. This behaviour differs from that reported previously for the photorefractive effect of FLCs under an applied DC electric field, wherein the gain coefficient decreased with the increasing strength of the external DC electric field. The refractive index grating formation time was measured based on the simplest single-carrier model for photorefractivity, whereby the gain transient is exponential. The grating formation time τ was obtained from the fitted curve.

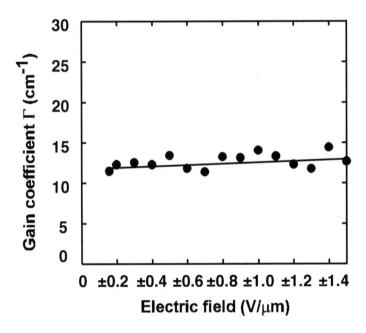

Figure 27. Gain coefficient of the two-beam coupling as a function of the applied AC electric field strength. The frequency of the field was 100 Hz.

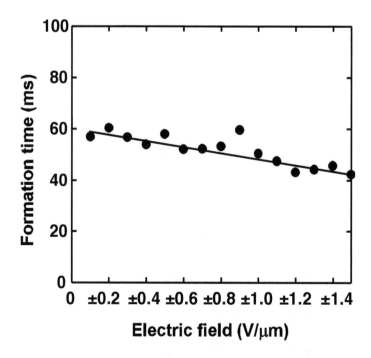

Figure 28. Grating formation time as a function of the AC electric field strength. The frequency of the AC field was 100 Hz.

As seen in Figure 28, the grating formation time decreased with an increased applied electric field strength as a result of the increased efficiency of charge generation. The grating formation time was determined to be 30-40 ms in the present case.

5.2. Frequency dependence of the gain coefficient and the grating formation time

The gain coefficients are plotted as a function of the frequency of the applied external electric field in Figure 29.

The gain coefficient was observed to increase with an increase in frequency within the range of 1 Hz to 100 Hz. The FLC exhibited switching based on polarization reversal at a frequency lower than 500 Hz. The gain coefficient reached a maximum at frequencies within the range of 60 Hz to 500 Hz. However, the magnitude of the gain coefficient decreased as the frequency was increased to higher than 500 Hz. This is thought to be because the FLC cannot perform a complete switching motion at frequencies higher than 500 Hz; therefore, the FLC molecules showed vibrational motion. Moreover, at frequencies from 500 Hz to 2000 Hz, some of the FLC molecules go through a rotational switching motion, whereas others undergo vibrational motion. When the frequency exceeded 2000 Hz, the magnitude of the gain coefficient reaches a constant value, regardless of the frequency. Thus, the value of the gain coefficient at frequencies lower than 500 Hz can be attributed to the photorefractive effect based on the complete switching of FLC molecules and, at frequencies higher than 2000 Hz, this represents the photorefractive effect based on the vibrational motion of FLC molecules caused by an alternating electric field. Figure 30 shows the frequency dependence of the grating formation time.

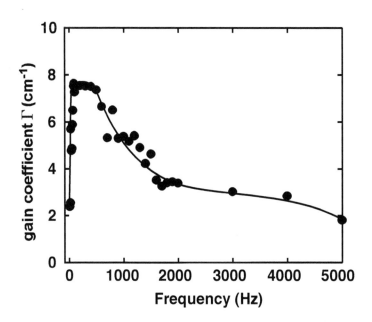

Figure 29. Gain coefficient as a function of the frequency of the applied electric field. The frequency was varied from 1 to 5000 Hz. The strength of the field was ± 0.5 V/μm.

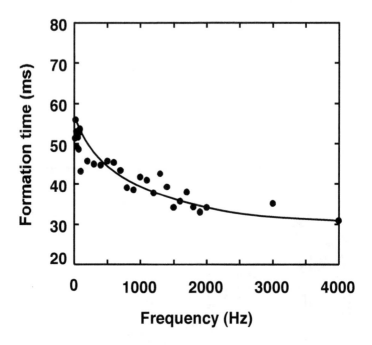

Figure 30. Grating formation time as a function of the frequency of the applied electric field. The frequency was varied from 1 Hz to 4000 Hz. The strength of the field was ± 0.5 V/μm.

The grating formation time was shortened as the frequency increased. In addition, the frequency dependence of the formation time was not correlated with the dependence of the gain coefficient. As the frequency of the applied electric field increased, the switching and vibrational motion of the FLC molecules is thought to be accelerated and the affect on the internal electric field becomes apparent. Asymmetric energy exchange in FLCs under a static DC field is dominated by the direction of the DC field. The increase/decrease in the beam intensity switches when the direction of the field is reversed. This originates from the direction of charge separation. Thus, an internal electric field is thought to be difficult to form under an AC electric field. However, asymmetric energy exchange was observed with good reproducibility, suggesting that an internal electric field is formed even under an applied AC electric field. In addition, no energy exchange was observed in any FLC lacking a photoconductive dopant under an AC electric field (Figure 31). This observation indicates that photoconductivity is necessary for two-beam coupling under an AC field.

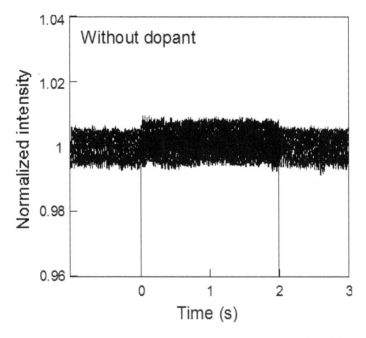

Figure 31. An example of the results of the two-beam coupling experiment under an AC electric field for SCE8 without a photoconductive dopant. An AC electric field of ± 0.5 V/μm, 100 Hz was applied. The shutter was opened at t = 0 s and closed at t = 2 s.

It is necessary to consider both hole transport by a hopping mechanism as well as ionic conduction in order to explain the formation of the space-charge field under an AC field. In addition, a number of experimental results support the large contribution of ionic conduction to the FLC medium. In addition, the photorefractive effect in SCE8 doped with 2 wt% ECz under an AC field was very small. If ionic conduction is the major contributor to the formation of the space-charge field, the anisotropic mobility of ionic species in the LC medium may affect the formation of the field. As shown in Figure 3, interference fringes are formed across the smectic layer so that migration of ionic species occurs in the inter-layer direction. The asymmetric structure of the surface stabilized state of the FLC may lead to the asymmetric mobility of cations and anions, which is thought to be one possible model explaining the charge separation under an AC electric field. However, the mechanism of the formation of the space-charge field under an AC field requires further investigation.

5.3. Two beam coupling experiment with the application of a biased AC field

An asymmetric energy exchange was observed upon the application of an AC field. This grating was interpreted as being based on the spatial difference in the molecular motion of the FLC molecules. The response time was on the order of a few tens of milliseconds and was dominated by the formation of the internal electric field. The photorefractivity in FLCs is accomplished through charge generation and diffusion. The application of an AC field to an FLC results in a very stable photorefractive response. However, it is obvious that an AC field is not advantageous for charge separation. Thus, the effect of a biased AC electric field on the photorefractivity of SS-FLCs was investigated. Figure 32 shows the concept of applying a biased AC field.

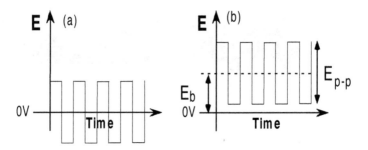

Figure 32. Concept of a biased AC field application. (a) Alternating electric field. (b) Biased alternating electric field.

A typical example of the two-beam coupling signal is observed with the application of a 2.0 $V_{p-p}\mu m^{-1}$, 100 Hz square waveform electric field without a bias field ($E_b = 0$), as shown in Figure 33(a). The transmitted intensities of the laser beams through the FLC/CDH/TNF mixture upon the application of an alternating electric field as a function of time is shown in the figure. The interference of the divided beams in the sample resulted in the increased transmittance of one of the beams and the decreased transmittance of the other. Although the transmitted intensity of the laser beam oscillates due to the switching motion of the FLC molecules, the average intensities of the beams were symmetrically changed, as shown in Figure 33(a). This indicates that a grating based on the spatial difference in the rotational switching motion of FLC molecules was formed and acted as a diffraction grating.

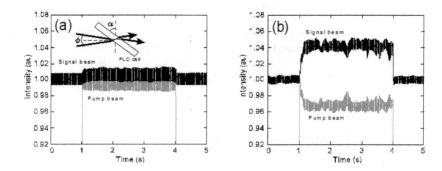

Figure 33. FigAsymmetric energy exchange with a square waveform of 2.0 $V_{p-p}\mu m^{-1}$, 100 Hz (a) without bias, and (b) with a bias of 0.5 V μm^{-1}. The intersection angle ϕ was 20° and the sample angle a was 50°.

The asymmetric energy exchange between the signal and the pump beams proves the formation of a diffraction grating phase-shifted from the interference fringe. The grating is considered to be based on the spatial difference in the rotational switching motion of FLC molecules. When the biased electric field was set to $E_b = 0.5$ V μm^{-1}, the two-beam coupling signal was dramatically enhanced (Figure 33(b)). The gain coefficient increased from 10 cm^{-1} ($E_b = 0$) to 70 cm^{-1} ($E_b = 0.5$ V μm^{-1}). Obviously, the biased field contributes to the formation of the diffraction grating. The internal electric field is thought to be difficult to form under an AC electric field. It has been reported that ionic conduction is the major contributor to the formation of the space-charge field in the photorefractive effect of an FLC. The anisotropic mobility of ionic species in an LC medium affects the formation of the field. In the experimental conditions reported, the interference fringe is formed across the smectic layer so that migration of ionic species occurs in the inter-layer direction. The asymmetric structure of the surface stabilized state of the FLC may lead to the asymmetric mobility of cations and anions. It was considered that the charge separated state is more effectively formed under a biased AC field.

6. Conclusion

The reorientational photorefractive effect based on the response of bulk polarization was observed in dye-doped FLC samples. Photorefractivity was observed only in the ferroelectric phase of these samples and the refractive index formation time was found to be shorter than that of nematic LCs. The response time was in the order of ms and was dominated by the formation of the internal electric field. These results indicate that the mechanism responsible for refractive index grating formation in FLCs is different from that for non-ferroelectric materials and that it is clearly related to the ferroelectric properties of the material. The photorefractivity of FLCs was strongly affected by the properties of the FLCs themselves. Besides, with properties such as spontaneous polarization, viscosity and phase transition temperature, the homogeneity of the SS-state was also found to be a major factor. The gain coefficient, refractive index grating formation time (response time) and stability of the two-beam coupling signal were all strongly affected by the homogeneity of the SS-state. Therefore, a highly homogeneous SS-state is necessary to create a photorefractive device. The techniques employed recently in the development of fine LC display panels will be utilized in the future in the fabrication of photorefractive devices.

Acknowledgment

The authors would like to thank the Japan Science and Technology Agency (JST) S-innovation and the Canon Foundation for support.

Author details

Takeo Sasaki

Tokyo University of Science, Japan

References

[1] Yeh, P. Introduction to Photorefractive Nonlinear Optics; John Wiley: New York, 1993.

[2] Solymar, L.; Webb, J. D.; Grunnet-Jepsen, A. The Physics and Applications of Photorefractive Materials; Oxford: New York, 1996.

[3] Moerner, W. E.; Silence, S. M. Chem. Rev. 1994, 94, 127-155.

[4] Meerholz, K.; Volodin, B. L.; Kippelen, B.; Peyghambarian, N. Nature 1994, 371, 497-500.

[5] Kippelen, B.; Marder, S. R.; Hendrickx, E.; Maldonado, J. L.; Guillemet, G.; Volodin, B. L.; Steele, D. D.; Enami, Y.; Sandalphon; Yao, Y. J.; Wang, J. F.; Röckel, H.; Erskine, L.; Peyghambarian, N. Science 1998, 279, 54-56.

[6] Kippelen, B.; Peyghambarian, N. Advances in Polymer Science, Polymers for Photonics Applications II, Springer, 2002, 87-156.

[7] Ostroverkhova, O.; Moerner, W. E. Chem. Rev., 2004, 104, 3267-3314.

[8] Sasaki, T. Polymer Journal, 2005, 37, 797-812.

[9] Tay, S.; Blanche, P. A.; Voorakaranam, R.; Tunc, A. V.; Lin, W.; Rokutanda, S.; Gu, T.; Flores, D.; Wang, P.; Li, G.; Hilarie, P.; Thomas, J.; Norwood, R. A.; Yamamoto, M.; Peyghambarian, N. Nature, 2008, 451, 694-698.

[10] Blanche, P. A.; Bablumian, A.; Voorakaranam, R.; Christenson, C.; Lin, W.; Gu, T.; Flores, D.; Wang, P.; Hsieh, W. Y.; Kathaperumal, M.; Rachwal, B.; Siddiqui, O.; Thomas, J.; Norwood, R. A.; Yamamoto, M.; Peyghambarian, N. Nature, 2010, 468, 80-83.

[11] Wiederrecht, G. P., Yoon, B. A., Wasielewski, M. R. Adv. Materials 2000, 12, 1533-1536.

[12] Sasaki, T.; Katsuragi, A.; Mochizuki, O.; Nakazawa, Y. J. Phys. Chem. B, 2003, 107, 7659-7665.

[13] Talarico, M.; Goelemme, A. Nature Mater., 2006, 5, 185-188.

[14] Sasaki, T; Miyazaki, D.; Akaike, K.; Ikegami, M.; Naka, Y. J. Mater. Chem., 2011, 21, 8678-8686.

[15] Skarp, K.; Handschy, M. A. Mol. Cryst. Liq. Cryst. 1988, 165, 439-569.

[16] Sasaki, T.; Mochizuki, O.; Nakazawa, N.; Fukunaga, G.; Nakamura, T.; Noborio, K. Appl. Phys. Lett., 2004, 85, 1329-1331.

[17] Sasaki, T.; Mochizuki, O.; Nakazawa, Y.; Noborio, K. J. Phys. Chem. B, 2004, 108, 17083-17088.

Permissions

The contributors of this book come from diverse backgrounds, making this book a truly international effort. This book will bring forth new frontiers with its revolutionizing research information and detailed analysis of the nascent developments around the world.

We would like to thank Dr. Aimé Peláiz Barranco, for lending her expertise to make the book truly unique. She has played a crucial role in the development of this book. Without her invaluable contribution this book wouldn't have been possible. She has made vital efforts to compile up to date information on the varied aspects of this subject to make this book a valuable addition to the collection of many professionals and students.

This book was conceptualized with the vision of imparting up-to-date information and advanced data in this field. To ensure the same, a matchless editorial board was set up. Every individual on the board went through rigorous rounds of assessment to prove their worth. After which they invested a large part of their time researching and compiling the most relevant data for our readers. Conferences and sessions were held from time to time between the editorial board and the contributing authors to present the data in the most comprehensible form. The editorial team has worked tirelessly to provide valuable and valid information to help people across the globe.

Every chapter published in this book has been scrutinized by our experts. Their significance has been extensively debated. The topics covered herein carry significant findings which will fuel the growth of the discipline. They may even be implemented as practical applications or may be referred to as a beginning point for another development. Chapters in this book were first published by InTech; hereby published with permission under the Creative Commons Attribution License or equivalent.

The editorial board has been involved in producing this book since its inception. They have spent rigorous hours researching and exploring the diverse topics which have resulted in the successful publishing of this book. They have passed on their knowledge of decades through this book. To expedite this challenging task, the publisher supported the team at every step. A small team of assistant editors was also appointed to further simplify the editing procedure and attain best results for the readers.

Our editorial team has been hand-picked from every corner of the world. Their multi-ethnicity adds dynamic inputs to the discussions which result in innovative outcomes. These outcomes are then further discussed with the researchers and contributors who give their valuable feedback and opinion regarding the same. The feedback is then collaborated with the researches and they are edited in a comprehensive manner to aid the understanding of the subject.

Apart from the editorial board, the designing team has also invested a significant amount of their time in understanding the subject and creating the most relevant covers. They scrutinized every image to scout for the most suitable representation of the subject and create an appropriate cover for the book.

The publishing team has been involved in this book since its early stages. They were actively engaged in every process, be it collecting the data, connecting with the contributors or procuring relevant information. The team has been an ardent support to the editorial, designing and production team. Their endless efforts to recruit the best for this project, has resulted in the accomplishment of this book. They are a veteran in the field of academics and their pool of knowledge is as vast as their experience in printing. Their expertise and guidance has proved useful at every step. Their uncompromising quality standards have made this book an exceptional effort. Their encouragement from time to time has been an inspiration for everyone.

The publisher and the editorial board hope that this book will prove to be a valuable piece of knowledge for researchers, students, practitioners and scholars across the globe.

List of Contributors

Hiroki Taniguchi and Mitsuru Itoh
Materials and Science Laboratory, Tokyo Institute of Technology, Yokohama, Japan

Hiroki Moriwake
Nanostructures Research Laboratory, Japan Fine Ceramics Center, Nagoya, Japan

Toshirou Yagi
Research Institute for Electronic Science, Hokkaido University, Sapporo, Japan

Wan Qiang
Institute of Structural Mechanics, China Academy of Engineering Physics, Mianyang, Sichuan, China

Chen Changqing and Shen Yapeng
Key Laboratory, School of Astronautics and Aeronautics, Xi'an Jiaotong University, Xi'an, China

N. D. Scarisoreanu, R. Birjega, A. Andrei and M. Dinescu
NILPRP, National Institute for Laser, Plasma & Radiation Physics, Bucharest, Romania

F. Craciun
CNR-ISC, Istituto dei Sistemi Complessi, Area della Ricerca Roma-Tor Vergata, Roma, Italy

C. Galassi
CNR-ISTEC, Istituto di Scienza e Tecnologia dei Materiali Ceramici, Faenza, Italy

A. Suárez-Gómez
UdG-CUVALLES, Carr. Guadalajara-Ameca, Ameca, Jalisco, México

J.M. Saniger-Blesa
CCADET-UNAM, Cd. Universitaria, Coyoacán, México D.F., México

F. Calderón-Piñar
Fac. de Física/IMRE, San Lázaro y L, Univ. de la Habana, Habana, Cuba

Koukou Suu
Institute of Semiconductor and Electronics Technologies, ULVAC, Inc., Shizuoka, Japan

Matías Núñez
Consejo Nacional de Investigaciones Científicas y Técnicas (CONICET), Argentina Centro Atomico Bariloche, U-A Tecnologia de Materiales y Dispositivos, Division Materiales Nuclkheares, Argentina
Instituto de Ciencias Basicas, Universidad Nacional de Cuyo, Argentina Previously at North Carolina State University, Raeligh, North Carolina, USA

Jeffrey F. Webb
Faculty of Engineering, Computing and Science, Swinburne University of Technology, Sarawak Campus, Kuching, Sarawak, Malaysia

Kai-Huang Chen
Department of Electronics Engineering/Tung-Fang Design University/Taiwan, R.O.C., Taiwan

Chien-Min Cheng
Department of Electronic Engineering/Southern Taiwan University of Science and Technology/ Taiwan, R.O.C., Taiwan

Sean Wu
Department of Electronics Engineering/Tung-Fang Design University/Taiwan, R.O.C., Taiwan

Chin-Hsiung Liao and Jen-Hwan Tsai
Department of Mathematics and Physics/Chinese Air Force Academy/Taiwan, R.O.C., Taiwan

Yuriy Garbovskiy, Olena Zribi and Anatoliy Glushchenko
UCCS Center for the Biofrontiers Institute, Department of Physics, University of Colorado at Colorado Springs, Colorado Springs, Colorado, USA

Takeo Sasaki
Tokyo University of Science, Japan

Printed in the USA
CPSIA information can be obtained
at www.ICGtesting.com
JSHW011814301024
72690JS00002B/77